Climate Change and Political Action in Italy

Daniel Delatin Rodrigues

Climate Change and Political Action in Italy

palgrave
macmillan

Daniel Delatin Rodrigues
University of Milano-Bicocca
Milan, Italy

ISBN 978-3-031-98770-0 ISBN 978-3-031-98771-7 (eBook)
https://doi.org/10.1007/978-3-031-98771-7

This Palgrave Macmillan imprint is published by the registered company Springer Nature Switzerland AG
The registered company address is: Gewerbestrasse 11, 6330 Cham, Switzerland

Contents

CHAPTER 1

Introduction: Notes on the Practices of Counter-Extractivism

1.1 Introduction

Each chapter in this book addresses a theme closely tied to the climate and ecological crisis—land use, infrastructure, urban planning and common goods. These topics are not necessarily of universal interest, nor can they be adequately approached through a single, narrow analytical lens. The themes emerge as problematic from the fieldwork: they are inseparable from the concrete situations in which they manifested themselves as issues. They were not imposed by me on the subjects I followed—on the contrary, they derived from their practices and concerns. Obviously, other experiences were also encountered, but it was not my intention to offer a comprehensive overview of everything that happened. What follows is a decidedly partial account, grounded in equally partial encounters and situated engagements—yet rich in meaning.

Each of the situations selected—for reasons of empirical access and personal involvement—allows us to consider theoretical and political perspectives capable of illuminating the current critical condition. Whether in relation to the impact of the climate crisis on everyday life or the tensions caused by the immobility of political institutions, these situations make visible the categories with which active subjects try to make sense of their experiences. By situating these responses within material practices generated in environments structured by fossil fuel ubiquity, it

D. Delatin Rodrigues, *Climate Change and Political Action in Italy*, https://doi.org/10.1007/978-3-031-98771-7_1

becomes possible to compare them with other territorial experiences. This comparative effort, however, is restricted here to the case of Milan.

The aim is not to establish a hierarchy among responses, but to explore—following Pignarre and Stengers—the instances of subjective and collective elaboration in response to critical situations: the ways in which those involved define the problem and construct possible means of coping. Thus, the cases analysed here will be presented as networks of partial connections between humans and non-humans, in practical attempts at non-extractivist life; extractivism is taken here as a mode of organising life centred on the large-scale appropriation of natural and human resources through processes of subjugation, depletion, and non-reciprocity, often reinforcing socio-ecological destruction and power asymmetries (Chagnon et al. 2022). It is important to stress that partial connections are not synonymous with local connections: for example, I will analyse how agroecological practices developed in south-eastern Brazil by a Swiss agronomist were transplanted and reinterpreted on the southern outskirts of Milan, in conjunction with a heterogeneous collective of young activists, students, artists, immigrants and precarious workers.

This entire work is structured around the question: How do people respond to climate and ecological change? But the intention is not to record visions or representations of these crises—but rather precarious and unstable experiences that must survive in an hostile environment. That's why this is not a critique project in the traditional sense, whose aim would be to disqualify these practices in the name of a higher structural, historical or normative truth. Rather, it is an investigation that is sensitive to the relationship between practices and the socio-material contexts in which they emerge.

The material collected accompanies these different responses. The answers here are always situated—as Stengers proposes—because they refer to unique problem situations that can be empirically reconstructed using multiple techniques. The experiences of these subjects—and the way they formulate responses in specific contexts—are entangled in complex webs of materials, affections, infrastructures and knowledge. Even among those who define themselves as climate activists, it is possible to identify important divergences on how to tackle the climate crisis. Rather than talking about ideological divides, it seems more productive to point to differences in the instances of problematisation: some give primacy to technical-scientific knowledge, or institutional and international politics, as universal criteria for evaluation; others value locally rooted forms of knowledge and practices, and their own ways of constructing meaning.

The conflicts are not limited to the dichotomy between denialism and acceptance of climate change. Various spokespeople for the event—physicists, engineers, architects and politicians—try to propose generalisable answers. The variety of responses is often treated as a relativism to be corrected. This book proposes a different path: following how those who become active seek to simultaneously escape fossil sociality and invent new existential territories (Guattari 2013), experimenting with alternatives in the field of mobility, food, the use of urban space and energy.

Milan is taken here as a microcosm to observe new practices for dealing with the climate crisis. Today, the city is witness to a multiplicity of experimentation, resistance, conflicts over habitability, attempts at desertion and forms of action that seek to sabotage extractives continuity. The practices analysed here emerged at a relatively favourable time for experimentation, which is not necessarily sustained in the present: today, many of the actions of these activists would be classified as a crime and punishable by up to three years in prison. This justifies the decision to keep the identity of most of those involved anonymous.

The central object of this book is experience. More specifically, it is about empirically accompanying what it means to live in a time of climate and ecological crises—and experimenting, from this, with situated responses that present themselves as prefiguration's (albeit fragile and unfinished) of non-extractives ways of life. The starting point was an open and undefined notion of activism: becoming active. This made it possible to include both subjects engaged in direct action and those who, for example, plant trees or experiment with agroecology. The experimentation front is broad. The common denominator among these subjects is the attempt to deactivate the extractive chains that sustain the acceleration of climate collapse—whether through direct action or the creation of agroforestry.

For this reason, I decided to work with a deliberately broad and imprecise notion of politics: here understood as the endeavour to build lasting ties with other humans—and also with forms of life and agencies that exceed the human. It makes no sense to judge the practices analysed here based on grand normative theories that aim to dictate what is right or wrong. What is at stake are ways of making worlds (Haraway 2016), that is, of fabricating forms of existence that try to deactivate dependence on fossil fuels and extractive chains. These practices involve real difficulties, practical limits and uncertain horizons for action—but they are concrete expressions of attempts to reconfigure collective life, often formulated in a smaller and dispersed key, in different territories.

These struggles don't fit easily into the grand revolutionary narratives that have shaped the imaginary of systemic transformations. As far as climate change is concerned, there is no shortage of information: for decades, the Intergovernmental Panel on Climate Change (IPCC) has produced detailed and consistent diagnoses of the problem, and the same goes for the wider ecological crisis, especially the collapse of biodiversity. Despite this, it is clear that this data does not translate unmediated into what would be an emergency for large-scale and lasting political action. As Danowski and Viveiros de Castro (2014) observe, different ethnographic records already document the varied ways in which human cultures have imagined the disarticulation of space–time structures. Climate collapse destabilises nature-culture entanglements and imposes a double challenge on those who feel affected: avoiding the worst and learning to live in a radically unpredictable world.

The image of the "emergency brake" evoked by Walter Benjamin takes on renewed relevance here: this book deals precisely with these brake practices—the actions that seek to interrupt extractivist continuity. But the focus is not on describing these practices as simple reactions to global capitalism, nor on seeing them as folkloric remnants of a more or less recent past. Rather, it is to record how these practices engender systematic alternatives, even if partial and precarious, in the midst of networks in which extractivism dominates personal and collective ways of life. The question that remains open is whether or not these worlds can acquire consistency and proliferate. Good intentions, as we know, are not enough. The survival of these practices depends on concrete conditions that allow them to persist in an environment that, in many cases, is openly hostile to their emergence.

My own experience—of disquiet and estrangement—became part of the research process. The dissociation between climate concerns and everyday action became acute. The gap was not so much cognitive as practical: the material and immaterial conditions that would allow for non-extractives forms of reproduction were lacking (or precarious). There was no attempt, however, to turn this research into an exercise in autoethnography. But the personal experience served as a starting point for a fundamental question: How can we act in a way that is coherent with the current state of the planet? The answer, as so many have experienced, is far from obvious.

So, I started observing what was happening in Milan. From the start, it was clear that my focus would not be on studying governance—not because I wasn't interested, but because my starting point was different: I wanted to know what people were doing about the climate crisis. How they reacted, how they imagined the future, what was changing in their lives, and based on what practices and alliances. I tried to adapt the tools of analysis to the concrete forms of engagement of those who were becoming active—with their hesitations, bifurcations and inventions.

I've been fortunate enough to start this investigation at a decisive moment for the international climate movement. Although we can't measure its future consequences, the years 2018 and 2019 will remain a milestone. In Milan, something was also happening. I followed dozens of events, demonstrations and meetings. I talked to participants, collected leaflets, recorded practices, listened to narratives and tried to understand the objectives—explicit and implicit—of these gestures. The diversity of groups, formats and themes presented me with a considerable analytical challenge. How could I maintain the relevance of these experiences if not all of them fell into the classic categories of activism, collective action or social movements?

The solution was pragmatic: to take becoming active as a starting point—a notion that is purposefully broad and crossed by diverse experiences. This allowed me to include everything from activists engaged in direct action to people involved in agroecology, reforestation or critical consumption. And because I took experience as the guiding thread, the focus fell on what actually happens when someone decides to do something—and with whom or with what this action is done. This was decisive: it wasn't a question of defining a type of engagement in advance, but of following how forms of participation and involvement emerge from problematic situations and gain meaning from the agencies—human and non-human—that make them up.

If, at first glance, the direction adopted seemed to take me away from urban studies, it was actually a question of explaining how urban materiality becomes problematic, in the sense attributed to the term by pragmatist philosopher John Dewey (2016): the subjects who become active and try to respond to climate change do so in particular conditions—institutional, material and subjective—that offer resistance. They find themselves "trapped" in the fossilised reproduction of business-as-usual. Fossil fuels don't just circulate through the city's cables and pipelines, or through the engines of cars and household appliances: they also sustain forms of life—fossil socialities (a theme that will be explored in greater depth in Chapter 2)—in other words, practices of personal and collective production and reproduction that are generated and maintained by this energy infrastructure.

This is how the minimum conditions for my fieldwork were established. My initial intention was to carry out participant observation of these experiences in order to understand how they materialised and, at the same time, how the urban infrastructure became an obstacle or a field of tension for those seeking to intervene. However, the Covid-19 health emergency removed many possibilities for face-to-face empirical research, forcing me to resort to other sources of data, digital platforms and alternative forms of listening. Unfortunately, not all aspects of face-to-face observation—such as gestures, silences and atmospheres—can be replaced by online research. Even so, between 2018 and 2019 I had the opportunity to take part in dozens of events, which allowed me to observe not only the content discussed, but also the ways in which this content was enunciated, performed and circulated.

I therefore had to diversify my research strategies. My face-to-face observations began in December 2018, at one of the first demonstrations in Piazza della Scala organised by the local Fridays for Future group, and continued—with intermittencies imposed by the pandemic—until 2021, during the pre-COP26 in Milan. The field included street demonstrations, open assemblies, collective meetings, public conferences and, above all, practical activities such as workshops, community gardens and volunteer work in agroforestry and urban gardens. At various times, the roles of researcher and activist were inevitably and productively intertwined.

Climate change and ecological collapse are today central challenges for contemporary societies—not only on a scientific and technical level, but also in the political, ethical and everyday spheres. This book sets out to investigate the forms of political action in response to the climate crisis from a situated study: the case of Milan between 2019 and 2021, a city marked by overlapping processes of environmental degradation, urban transformation, the emergence of new climate citizenships and experimentation with non-fossil socialities.

The central aim is to offer a relational and practice-orientated account of how political action emerges, transforms and confronts structural blockages in urban contexts. Rather than privileging formal institutions or large environmental movements, this work focuses on the active becoming of individuals and collectives: it investigates how common practices, material infrastructures and everyday ecologies make up the concrete terrain of climate politics.

The notion of fossil socialities is central to the argument: it is about articulating how infrastructures, technologies and daily routines sustain the reproduction of the extractivist way of life and, at the same time, limit the possibility of ecological transformation. In contrast, the book explores practices of rupture, care and regeneration that seek to destabilise these continuities.

In this context, becoming active is not understood as mere ideological adherence, but as a performative, precarious and materially situated process. This process encompasses a broad spectrum of actions—from mass protests to agroecological practices, from occupying public spaces to reconfiguring domestic consumption habits. These practices challenge traditional conceptions of activism and reveal how climate politics emerges from everyday engagement with socio-material networks, bringing together humans and non-humans on multiple scales.

This book is structured to reflect the complexity and partiality of the experiences analysed. Each chapter focuses on a specific dimension of climate action in Milan—from resilience policies and green infrastructure to territorial conflicts and regenerative practices. Although the narrative takes on an impressionistic character, made up of fragments and localised situations, it is guided by a consistent analytical thread: understanding how political practices aimed at the climate crisis are generated, sustained and sometimes tensioned or interrupted in an urban environment whose limitations to transformations cannot be disregarded.

By accompanying these situated responses, the book seeks not only to document emerging forms of climate activism, but also to contribute to the reimagining of what political action can be in the Anthropocene. It invites the reader to go beyond reductive binarisms—between individual and systemic change, between the local and the global, between human and non-human agencies—and instead proposes a more interwoven, relational and dynamic perspective on political engagement in times of ecological collapse.

Ultimately, the aim is to offer a critical—but not cynical—exploration of the possibilities of building fairer and more liveable futures from everyday acts of resistance, invention and care, even when these acts are partial, precarious and provisional. I tried to maintain a narrative form in which it would be possible to follow, in the subjects' own words, how they tried to respond to the climate crisis and what obstacles they faced along the way.

1.2 Organisation of the Book

The contribution of this book lies in the articulation between three interdependent axes: a practical-material approach to environmental engagement; an analysis of urban political ecologies under regimes of institutional resilience; and an analytical proposal centred on the notion of becoming active as a contingent, relational and situated process. Based on a case study in Milan, the book seeks to understand how subjects and collectives respond to the climate crisis in urban contexts marked by fossil dependencies, territorial disputes and attempts to reinvent everyday life. Rather than applying preconceived models of green citizenship or environmental participation, the research proposes an ecology of practices that emerges from the concrete monitoring of the intertwining of bodies, infrastructures, objects and living beings—human and non-human—that constitute the ways of making the world in the Anthropocene.

The book is structured around the observation of specific problematic situations. These situations, as defined by Isabelle Stengers, are not analytical abstractions, but empirical compositions where different forms of existence are put to the test and where it becomes necessary to invent answers, even without guarantees of success. The methodological option was to follow the practices as they became relevant to the subjects who engaged in them. This meant avoiding both normative judgements (what should be done) and totalising structural explanations (why things are the way they are) in order to focus on the ways in which people try to do something in the face of the crisis.

The diversity of practices observed also required an analytical approach that was attentive to their materiality and their conditions of emergence. Some of these practices are orientated towards caring for degraded urban territories, threatened plants and ecological cycles, while others are dedicated to building energy, food and infrastructure alternatives that seek to break away—even partially—from extractive chains. In all cases, these are practices that problematise everyday life, reconfigure relationships and open up gaps in the fossilised consensus of business-as-usual.

The book proposes that contemporary climate politics, when seen from below and from within urban practices, does not present itself as a large deliberate project, but as a multiplicity of experiments that operate in a distributed way. It is these experiments—sometimes contradictory, precarious, transitory—that make up the field of climate action today. In this sense, the idea of becoming active makes it possible to name this dynamic of partial activation that does not depend on a prior militant identity, but emerges from the confrontation with the ecological and political conditions that sustain the reproduction of life in common.

Instead of proposing a rigid typology or an ideal model of participation, the book builds a cartography of possible responses to the crisis, situated in Milan's urban socio-materiality, but potentially comparable to other geographies. The analyses dialogue with a wide range of authors and perspectives—from the anthropology of infrastructure to the political theory of prefiguration, from multi-species studies to the sociology of transition—making up a mosaic that is intended to be open and useful to other research, territories and experiments. The aim is not to offer a total theory or a normative guide to action, but to contribute to the collective endeavour of thinking and practising politics in times of collapse.

In Chapter 2, I examine some of the analytical challenges that climate change imposes on the interpretative structures of the social sciences. Our aim is to construct categories that do not exclude more-than-human agencies, thus allowing them to be included in social analysis. In Chapter 3, I present some aspects of the "resilient city" that has emerged in recent years. I try to offer a reading key to accompany the formation of what I call "climate citizenship". In Chapter 4, I follow some experiences that seek to "record" the agentivity of trees and plants in order to activate socio-ecological and climatic functionalities. This analysis is based on the urban forest programme and the transformation of the domestic environment into an ecological housing module. In Chapter 5, I turn my attention to the "green common good" proposed by activists as a political counterpoint to resilient city initiatives. In this chapter, I will present the territorial struggles in defence of Milan's green areas. In Chapter 6, I analyse the formation and practices of the Milano per il clima network, which brought together dozens of organisations and associations active in debates about the climate crisis in the city. The analysis focuses on how these collectives defined their own practices and the challenges faced in producing collective action from the diversity of subjects involved. Chapter 7 continues to analyse the practices of climate activists, but focuses on what I call practices of disruption. I have selected specific situations in the political articulation of these groups that highlight the heterogeneity, divergences and conflicts around the practices considered legitimate and effective for transforming the business-as-usual of climate policies. In Chapter 8, I present two experiences of generative practices. My aim is to follow how the subjects gathered around these experiences seek to turn more-than-human agencies into a vector for subjective and political transformation, offering—not univocally—a language that seeks to escape not only fossilised assemblages, but also the specialised practices of managing so-called ecological services.

The book therefore invites us to follow these trajectories of engagement without demanding their purity or effectiveness, but recognising in them attempts to sensitively reconfigure the common world. If there is a basic argument that runs through the text, it is that responses to climate change should be understood as material practices that emerge from specific situations and, in doing so, call for new modes of coexistence, care and belonging. Becoming active in this context is not just a political choice, but a process of deep involvement with a changing world—and with the urgency of imagining others.

1.3 A Brief Note on Methodological Aspects and Data Collection

1.3.1 Methodology and Theoretical Framework

This book is the result of a qualitative ethnographic investigation carried out between December 2018 and November 2021, focusing on the city of Milan, Italy. The research combined participant observation, documentary analysis and digital ethnography, and was gradually adapted to the restrictions imposed by the Covid-19 pandemic from 2020 onwards. The fieldwork included participation in assemblies, workshops, demonstrations, urban forums and direct actions promoted by climate collectives, as well as continuous monitoring of their dynamics on digital platforms and social networks.

The choice of Milan as a case study is justified by its ambiguous condition: on the one hand, it is a metropolis that positions itself as a reference in the promotion of "green" policies and "urban resilience"; on the other, it is the scene of persistent territorial conflicts and forms of climate activism that actively challenge this dominant model. The cases and episodes analysed were selected for their ability to illustrate tensions between different modes of political engagement—from large public mobilisations to everyday decarbonisation micro-practices.

Much of the data presented here was collected through fieldwork and participant observation, respecting the contextual negotiations required for this type of research. Other information was obtained from social networks and group chats in which I actively participated. This methodological choice, although motivated by the limitations imposed by the health emergency, entailed significant ethical challenges. Although I introduced myself as a researcher at the start of the investigation, my initial intention was not to use the chats as a formal source of data. For this reason, I chose to omit the names of the participants—a decision that also extended to other research contexts. Considering that many interlocutors were involved in political confrontation and that in recent years specific laws have been passed in Italy to dissuade climate activism, it seemed essential to me to guarantee the safety and privacy of those who, with courage and creativity, continue to experiment with ways of becoming active in the midst of the crisis.

References

Chagnon, Christopher W.; Francesco Durante; Barry K. Gills; Sophia E. Hagolani-Albov; Sanna Hokkanen; Sohvi M. J. Kangasluoma; Heidi Konttinen Markus Kröger; William LaFleur; Ossi Ollinaho and Marketta P. S. Vuola. (2022) "From Extractivism to Global Extractivism: The Evolution of an Organizing Concept". *The Journal of Peasant Studies*.

Dewey, John. (2016) The public and its problem: an essay in political economy. Ohio: Ohio University Press.

Guattari, Félix. (2013) "Qu'est-ce que l'écosophie?" Textes (1985–1992) présentés et agencés par Stéphane Nadaud. Paris: Éditions Lignes.

Haraway, Donna. (2016) *Staying with the trouble*: making kin in the Chthulucene. Durham: Duke University Press.

Viveiros de Castro, Eduardo; Danowski, Déborah. (2014) *Há mundo por vir?* Ensaios sobre os medos e os fins. Florianópolis: Cultura e Barbárie/ISA.

CHAPTER 2

More-Than-Human Agencies and Social Research

> "The constructivist question I have asked is 'what makes these human beings, these producers of the knowledge we call 'experimental', become active? In other words, 'what is the uniqueness of the adventure in which they become engaged?' 'What matters to them?' What does success mean to them?" (Stengers 2011: 20)
>
> "For animals, as well as plants, there have never been individuals" (Gilbert et al. 2012: 336)
>
> "Similarly, I would venture to say that there is no identity of a practice independent of its environment. This in no way means that the identity of a practice can be derived from its environment. Thinking 'par le milieu' does not give power to the environment. The dogged work and research of ethologists to discover what kinds of relationships with their monkeys would be the right ones for those monkeys to learn, whatever they learn, is enough to support the idea that the question is not one of power but of involvement. Spinoza could tell us that we don't know what a practice is capable of becoming; what we know instead is that the very way we define, or approach, a practice is part of the environment that produces its ethos." (Stenger 2005: 187)

When we ask ourselves what kind of transformations are needed to slow down the climate and ecological crisis and establish new forms of terrestrial habitability, we cannot limit ourselves to the expectation of a

D. Delatin Rodrigues, *Climate Change and Political Action in Italy*,
https://doi.org/10.1007/978-3-031-98771-7_2

simple change in human "consciousness" in relation to a supposed "natural world"—as if it were enough to adjust secondary qualities (values, perceptions, feelings) to the primary qualities of physical reality (Latour 2014, 2017; Clark 2021; Viveiros de Castro 2014). Responding to Gaia's insistence, as proposed by Latour and Stengers, rather requires modifying material relations: transforming practices and environments in such a way as to make explicit the more-than-human agentivity and limits of current ecologies (Kohn 2021; De la Cadena 2015; Kirksey 2012).

It is in this sense that the notion of attachments, developed by Stepanoff (2024), becomes fundamental for thinking about and qualifying forms of bonding, entanglement and interdependence between humans and non-humans. The thesis developed here argues that analysing the material aspects of responses to climate change not only allows us to observe them from a different angle but also provides the conditions for a tactical engagement of social theory in times of ecological collapse. This engagement is not limited to highlighting human affections and values ("secondary qualities") but seeks to follow agentivities in an expanded relational field, also tracing the political practices—both of resistance (anti) and invention (alter)—that emerge from these dynamics.

It's not a question of knowing a priori whether these practices are correct or effective in the traditional sense. What matters is what effects they produce and whether, even partially, they manage to interrupt or reorient the long extractive chains that structure contemporary life.

A plant dying on the balcony of a house can be interpreted as a sign of "lack of care", but also as an indication of the effects of a heatwave. A plastic bottle is both "full of oil" and linked to the disposable economy; a car is, in fact, a "climate agent"; buying something through Amazon is strengthening logistical chains of dependence and externalising costs. These everyday gestures generate mistrust or ambivalence: How can we trust an activist who depends on the same circuits he denounces? This contradiction doesn't invalidate the action—it makes explicit the dilemmas constitutive of any attempt at transformation situated within fossil sociality.[1]

Materiality communicates continuously—it is not a neutral backdrop, but an active agent that structures possibilities. It thus becomes an object of investigation in everyday life, as its presence reveals its insertion in

[1] These are "fragments" of conversations held with different activists at different points in this investigation.

an environment that favours certain practices and inhibits others (Gabrys 2020). The physical and emotional effort to transform everyday life through non-fossilised and non-extractive forms of connection involves complex situations of sacrifice, anxiety and even personal and family conflicts. Sometimes this effort opens escape routes: moving to another city or country, lifestyle changes (such as becoming vegetarian or vegan), the decision to sell the car and not fly anymore and many others.

What is normally understood as an "individual response" must therefore be situated in an environment that gives it meaning—that is, that offers the material, affective and symbolic conditions for such a response to be possible. Paradoxically, it is often the very systems of mediation—such as infrastructure, logistics, income, available time—that allow action to appear as "individual", reinforcing the impression of independence rather than interdependence.

Classifying them as individual responses is only possible because they have been produced as such—because their conditions of possibility and the networks that support them remain invisible or disregarded. This doesn't invalidate them; it just makes it more urgent to recognise how socio-material arrangements actively participate in the form that action takes.[2]

Talking about individual action often serves to keep the figure of the disconnected subject alive. As Debaise and Stengers (2021) emphasise, nothing is more dependent than the modern individual. The environment in which this individual operates is not just a neutral backdrop for the action of autonomous agents: it is an active medium that provides support and co-produces agents, enabling their production and reproduction (Dutreuil 2020). This dependence is realised through diffuse and capillary networks—the same ones that Stépanoff (2024) describes as shaping and guiding the bonds of existence and support.

For this reason, in this work I choose to use the notion of person rather than individual, and I speak of personal change rather than individual change. What I was able to observe in my fieldwork does not concern the autonomy of isolated subjects, but rather the transformation of the techniques for producing and reproducing the body—and the ways in which this body is engaged in social, technical and energetic (fossil) apparatuses that support a specific form of habits and behaviours, to the detriment of

[2] And that it is not a universal invariant of the human being. Individual autonomy is only the effect of the chain of dependence.

others. Inspired by Stengers' proposal, I prefer to say that subjects act in a populated and composed environment, not as a fixed backdrop to human stories, but as a medium that interferes, reacts and participates:

"Endurance cannot be explained by any virtue of an individual that endures; it depends first and foremost on the dynamics of infection. Everything that manages to endure has managed to infect its environment in a way that is compatible with this endurance" (Stengers 2011: 182).

By investigating what it means to become active, I try to offer a starting point that repositions the relationship between part and whole—between people and the socio-material and economic system—in such a way as to make it possible to articulate this becoming active, both personal and collective, as a practice of rupture, disconnection and dissociation from fossil socio-materiality. Instead of starting from a structural reading of the world-system to judge the insufficiency of individual actions, I start from the concrete experience of modifying the links and infrastructures that shape everyday life. ("We are not interested in individual changes, we in political ecology believe that we must offer a systemic critique"—Collettivo Ecologia Politica Milano Statale—16/06/21.)[3] In this work, I propose to take the opposite route—not, of course, to redefine the world-system, but to start from what are usually classified as personal actions, in order to demonstrate that such actions are immediately collective (Lapoujade 2020). This is because they are inscribed in discursive and material agencies that structure everyday life and that can, through practices of rupture, spread and reconfigure fossil bonds—what, following Tsing (2015) and Stengers (2020b; 2021), I call chains of dependence.

The so-called geological turn (Maniglier 2020: 62) needs to be thought of pragmatically, in terms of its consequences and the possibilities it opens for thought and action. This shift forces us to reconsider our modes of sociological imagination in relation to the modes of imagination of the very subjects who become the object of our investigations. These two dimensions should not be confused: on the one hand, there is the conceptual imagination of our interlocutors; on the other hand, there is the critical reflexivity of sociological description, which must redistribute agents and agencies in a balanced relationship. The conceptual imagination of our collaborators can thus not only situate our empirical data but

[3] https://www.facebook.com/events/483224626279261?ref=newsfeed

help us to critically reposition our own analytical categories—this is the movement I intend to follow.

To do this, I take as my starting point the process of becoming active, which involves people, objects, technologies, living beings and practices. Becoming active is understood here as an emergent, relational and more-than-human performance (Tsing et al. 2024), which is not limited to isolated human agency.

The distinction between personal and collective, as mobilised in this description, is anchored in the definition of sociality (and not society) proposed by Marilyn Strathern: "people have the potential to relate and are at the same time always embedded in a matrix of relations with others" (Strathern 1988: 239). It is precisely this matrix of relations that I consider crucial for investigating problematic situations in which it becomes possible to understand the point of view of those who become active.

I therefore propose the concept of fossil socialities to designate a relational matrix—material and immaterial—whose support is rooted partially in fossil fuels that includes the modern ideal of individual autonomy (Charbonnier 2020), which is only possible through the extensive use of fossil energy infrastructures and resources (Mitchell 2011). The exercise of autonomy, in this sense, is supported by an opaque and extensive chain of energy extraction and circulation, typical of large-scale extractive economies—a chain that, as studies in the anthropology of infrastructure show, remains largely invisible, even though it is technically indispensable.

2.1 After "Becoming Active"

> "With the notion of 'availability' the signs that mark the world and that mark the subject are redistributed in a new way. Both are active and both are transformed by the availability of the other. Both are articulated by what the other 'makes him/her make'." (Despret 2004: 125)

An investigation into the forms of climate activism can start from different points, just as there are many aspects of everyday, political-institutional and economic life that can serve as a gateway to the subject. Each starting point requires attention to the concrete situations in which these practices are articulated, because it is in these situations that both their relevance and their specific modes of existence are defined. Climate change has become a vector of problematisation (Stengers 2021) that crosses collective life without, however, offering a common horizon of political unification (Latour 2018; Latour 2020; Danowski and Viveiros de Castro 2014; Tsing, various articles).

This is a central point. The political challenge enunciated by Gaia's insistence (Stengers 2015) calls for the ability to formulate appropriate responses. But how should these responses be generated? The proposal here is to investigate the processes of generating these responses based on specific empirical conditions, while rejecting the existence of a pre-established model that provides definitive solutions. At the very least, it can be said that new openings have been created for political imagination and experimentation. The magnitude of the crisis forces us to rethink politics beyond the institutional or antagonistic contours in which it is usually restricted.

The sociological work developed here is inspired by the pragmatic approach that Stengers (2020a, b: 79) proposes as appropriate for sociological research: "Et, en sociologie, cela significifierait que, s'ils veulent obtenir des réponses dignes de ce nom, les enquêteurs se devraient de ne poser que des questions rencontrant les préoccupations actives de celles et ceux qu'ils interrogent". In other words, the problems worth pursuing are those that concretely affect the subjects involved. The climate activists observed in this research are heirs to the "Great Acceleration", which produced a social environment largely indifferent (especially in the Global North) to the devastating consequences (largely in the global South) of mass consumption and the promise of infinite abundance offered by the market economy (Mitchell 2011; McNeill and Engelke 2016; Charbonnier 2020).

This is the starting point for describing the contemporary conditions for action in the era of the crisis of habitability. The climate and ecological crises manifest itself as a crisis of presence and reproduction: the habits forged during the "golden years" (Schulz 2020) of modernisation have come to be experienced as destructive, ethically unsustainable and even criminal—becoming the object of moral condemnation. Ordinary forms of sociality are revealed as modes of predation that jeopardise the conditions of existence of other forms of life, drastically reducing both the possibility of survival of other species and that of humans themselves.

2.2 Suspension: Two Very Instructive Speeches

When I started my research, I decided to contact a well-known professor of environmental engineering and science communicator at the Politecnico di Milano (Polimi) in order to understand how the climate issue was being dealt with in the city. He answered my call with a mixture of surprise and irritation: I was addressing a specialist with questions that could only be answered, according to him, on the basis of what qualified him as such—public policies, international agreements, national energy policy and statistical data. My questions, of course, were poorly formulated—they were open-ended, exploratory and sought only to probe the ground. But the failure of that interview gave me a first clue as to how the climate issue could be approached—and, above all, how I wouldn't want to approach it.

The expert's impatience with my lack of clarity revealed more than simple frustration: it indicated that the topic, although of interest to him, required—according to him—prior technical preparation. I should get better informed, study the international treaties, the IPCC reports and the national energy plans. That, according to him, was where the "climate issue" really lay. My hesitation was read as a sign of disqualification: I remember the moment when he asked me, surprised, where I was doing my doctorate and who my supervisor was—perhaps thinking of the work this supervisor would have to do to "get me back on track".

This experience resurfaced in all its density during one of my first field trips with activist groups. It was at a meeting of Resilient Gap—a student collective active in Polimi. When I commented that their proposals didn't take into account geopolitical conflicts, the class struggle and the structural domination of capital—which, for me, would require a politics of rupture—one of the activists (P.) answered me calmly. He explained that their practice didn't deny these dimensions, but that they acted on the basis of "big and small interventions", those that were accessible to them as students in a city like Milan, and within a national and European context that still allowed for legal, institutional and peaceful forms of political struggle. At that meeting, they discussed actions such as clean-up efforts, educational games on climate negotiations, workshops organised by the Milano per il Clima network, and awareness-raising events on environmental justice.

It was then that I realised something decisive: I ran the risk of repeating with the activists what had been done to me by the expert. If he had disqualified my research for not following the technical path he expected, I, in turn, was disqualifying the activists' practices for not corresponding to what I thought was politically necessary.

These two episodes became too important to ignore. The difficulty of unequivocally defining how climate change affects people and how people can or should respond to it—without ambiguity—led me to critically rethink not only my field approach, but also some common assumptions in climate research.

These differences—between scientific specialisation, political activism and everyday engagement practices—should not be treated as errors, but as constraints that shape the very course of research. The first lesson, therefore, is not to take these differences as proof of misunderstanding about what "really" matters. The second is to make them a guiding thread: a way of following what is required by each position, what it mobilises to legitimise and propagate itself, and how it can be transformed in the encounter with others. The first is not to make these differences the demonstration of a mistake about what matters; the second is to make these differences a leitmotif that allows us to follow "what is required by the position whose legitimacy it expresses, what it recruits to support or propagate, and how it is likely to be affected by the encounter with another position" (Stengers 2011: 513).

In urban contexts—as in other areas—it is a question of articulating new forms of encounter and coexistence. Evoking the exteriority of carbon and the ubiquity of capital does not authorise us to ignore the fact that those who need to respond to these two forms of transcendence (Stengers 2015a) always do so from concrete situations, facing specific problems that coexist with issues raised in other instances and places—or that are imposed on them. Furthermore, these responses are translated from and through the situations and connections of which these people are a part.

2.3 Experiencing Problematic Situations

> "What is required by your hold? From what wager does your success proceed?" (Stengers 2011: 518)

But this experience doesn't just concern the object of the research (activists, experts), it also involves the researcher himself. This is what makes the difference between investigating a problem and investigating problematic situations and forms of problematisation—the definition of which must be pertinent not only to the researcher, but above all to those they are following. "Thinking with those who become active" means starting from the concrete experience of those who recognise themselves as activists and seeking to elaborate, conceptually, not only a redefinition of their practices—and of what can be included in the category of activist—but also of the problematic field to which their activity refers and without which activism would have no object of intervention. What situations does this activism respond to?

These problematic situations, however, are not the same for the activist, the environmental engineer or the researcher. Recognising this difference is one of the aspects that emerges from empirical research into how to tackle and respond to climate change—and brings us directly back to the crisis of abstraction pointed out by Debaise (2017). It is a question of disorientation, not only as to what to do, but also as to how to do it, in the face of something that is almost always mediated by technologies, materialities, non-human agents or extreme climatic events (heatwaves, droughts, storms, etc.).

It is in these smaller, marginalised forms that we can also follow the emergence of responses to climate change. But for this to happen, such responses need to be situated in environments that offer the means and language through which they can be codified, interpreted and problematised. It's no surprise, then, that some of these solutions to fossil socialities take on characteristics that evoke a *déjà vu* of past experiences—from the counterculture to current movements such as slow food. Experiences previously supplanted by the modernisation front (Latour 2013) are reactivated under a new key that makes them contemporary, rescuing traditions to offer pertinent responses to the present.

However, it's not just a question of describing them as anachronistic. These subjects inhabit the interstices of the fossil urban machine and the ruins of capitalist modernisation. The alternatives they propose emerge from marginalised territories or commercial circuits that manage to remain, even partially, outside the chains of fossil dependency. Many of the experiences analysed in this book are in abandoned or peripheral spaces. Learning to inhabit these territories implies building other references that do not exclude them from value or the possibility of becoming vectors for experimentation—beyond fossil sociality and cemented urbanisation.

This experimentation can be understood as a practice of bricolage (Viveiros de Castro 2019), in the sense that it relies on the reinvention of means and instruments from available materials. However, this does not cancel out its originality: the operation carried out consists precisely of proposing solutions to the problem of a "social" to be rebuilt by subtracting the elements that threaten the conditions of habitability—personal and collective, human and non-human. The response to climate change translates, in practice, into a technical apprenticeship aimed at dismantling the "Anthropocene machine".

Autonomy, as Charbonnier (2020) argues, presupposes the possibility for subjects to free themselves from the constraints imposed by their environment. But this same autonomy is an effect of the fossil infrastructure. The abundance that is desired is that promoted by the value of unlimited growth within an environment that should ideally offer infinite opportunities—at the cost of sacrificing distant territories. The activist who turns to the local fossil infrastructure confronts the chains of dependency that operate on a planetary scale. They will confront the interconnected mechanisms that allow extractivism to function as a system, structuring individual and collective choices and possibilities.

The "system to be changed" therefore needs to be redefined in the light of this concrete, every day and immediate dimension, which is both captured and tensioned by disruptive practices. We can thus welcome the conceptualisations of the activists themselves and, at the same time, recover a systematic definition of these practices—without falling back on the dichotomous opposition that often marks the theoretical debate on climate action.

2.4 What Can an Activist Do?

This first contact with activists and experts allowed me to reformulate the initial research hypothesis: climate activism cannot be analysed on the basis of a fixed set of attributes. Instead, I start from what the activists themselves consider pertinent—the practices and problems from which they seek to formulate responses. These responses translate the climate event into situational and concrete terms, changing habits, transforming urban spaces, occupying the streets with protests and artistic performances, planting trees, promoting exchanges of second-hand goods, teaching each other how to build low-carbon lifestyles. In short, it's about learning to live—trying, step by step, to escape the addictive modes of existence that make up fossilised life.

Looking at the diversity of experiences, collective subjects and associations that make up the "Milano per il Clima" network, it is possible to draw some lessons. These do not invalidate the ways in which these individuals justify their political engagement, but they do allow us to identify the difficulties inherent in producing a common world. Instead of assuming an automatic unification of interests in the face of a global threat, what we see is a multiplicity of trajectories. Becoming active does not correspond to a single form of engagement but is entirely situational—or even multi-situational—and therefore heterogeneous. This heterogeneity should not be dismissed as a mistake in the face of structural analyses, because it emerges precisely from the translation of global climate change into the practical and material situations in which subjects are immersed. In other words, from the experiences produced by the infrastructures and objects of fossil life.

It is from the experience of these environments that activists problematise dynamics that, although rooted locally, refer to extra-local scales. But how can this experience be traced? Below, I present some personal communications collected during the field research, which show how climate politics should also be observed through an active becoming of objects and materials—capable of provoking, of "getting other agents to do something":

> "I live day to day. I know the climate situation is serious, but I don't think about it all the time. It worries me, but I don't know exactly what to do. When I go to the kitchen to get water, I find the climate issue on the plastic bottle. It makes me think. I feel guilty. I'm looking for solutions to install a filter that turns tap water into sparkling water. Then I wouldn't have to buy all those plastic bottles" (L., Anthropologist, 37 years old, January 2020).
>
> "I contribute to emissions. Sitting at the computer contributes to emissions. Now I cycle and no longer use plastic bottles. But the problem isn't just individual. If we don't confront the power of the oil industry, it won't be enough to change our lifestyles" (M., Geographer, September 2019).
>
> "Today I had to buy an iPhone for my daughter. Then, of course, the question of product disposal came to mind, the short lifespan, all the material and human costs. But what was I supposed to do? She broke her mobile phone. And she asked me for another one" (F., Anthropologist, October 2020).

> "I always assess the impact; how it was produced, whether it used oil, whether it used chemicals. All this is fundamental; there's always a climate dimension that can be discovered" (A., student, 26 years old).

Stories like this could be repeated ad infinitum. It would be difficult not to recognise, in the everyday practices shaped by extractive chains, the mechanisms that are exacerbating the climate and ecological crisis. They all refer to "things"—objects, resources, technologies—and to ordinary practices such as drinking water or choosing what to wear. They also refer to interpersonal and affective relationships: "should I buy this for my child?", "is it ethical to continue using this type of product?". In this way, mundane aspects of everyday life are placed in a new light, becoming problematic. Objects that once symbolised progress are now likely to be labelled "objects of the Anthropocene" (Chakrabarty 2020).

More than that, these situations reveal the persistence of a business-as-usual sustained not only by individual choices, but also by an intricate set of socio-technical, spatial and ideological mechanisms. This daily reproduction of fossil life depends on infrastructures whose lifespan endure on time, urban planning decisions made decades ago that today restrict the possibilities for experimentation, amenities guaranteed by the intensive use of fossil fuels and, of course, the oil and gas industrial complex (Grasso 2022), which controls energy sources, defines investment plans and contaminates the social environment with the promise of affordable goods and unlimited economic growth.

This context requires us to rethink the question of power and political action. Unlike the direct actions carried out by trade unions at the height—and during the decline—of the carbon economy (Mitchell 2011; Malm 2016), the long distance from the production and circulation networks of fossil fuels imposes other modes of confrontation. I'm not proposing to present a programme or outline a political (or tactical) strategy here. In various parts of the world, actions have already been taken to block the expansion of extractive operations and deactivate those in operation. It is on this empirical and conceptual basis that I propose to outline a theory of deactivation—not with the intention of universalising it, but as a provisional attempt at synthesis within the analytical and disciplinary limits that I will explain throughout the book.

This theory, it is worth emphasising, is not concentrated in a specific section, but is distributed transversally throughout the book. It is through this theory that I seek to understand how concerns about everyday practices—such as consumption or mobility—can become occasions for transforming the material bonds around us, revealing points of vulnerability and opening gaps for other forms of existence.

2.5 Activation Performances

With climate change, we are witnessing an increase in non-human agentivity, which manifests itself through bodily devices, technologies and techniques (Hawkins 2011; Marres 2012; Gabrys 2014). As I will show throughout this book, urban spaces are privileged crossing points where chains of dependency and interdependent relationships intertwine and overlap.

The responses I'm analysing here should be understood as versions that put into practice—in the sense of experimental science practices—what I will call activation performances. These performances are partial and situated: they depend on the present conditions and can be activated or deactivated depending on the context. In certain scenarios, they prove to be pertinent; in others, they become innocuous or incapable of mobilising an interlocutor or effect. Problematic situations create opportunities for performances in which materiality that was previously inert or indifferent—in other words, unable to respond or introduce a difference—starts to act as a vehicle for problematisation ("what is the ecological cost of a plastic bottle?") and engagement ("I need to change the way I consume").

These performances of activation articulate the construction of subjectivities, material politics and collective practices, cutting across different forms of attachment and putting the question of material participation back at the centre of analysis (Marres 2015). Based on this concept, it is possible to observe that the responses differ based on two main vectors: on the one hand, a contraction vector, which tends to ignore the interconnections between ways of life and focus on local and extra-local aspects defined as central to the fight against climate change (carbon emissions; international agreements); on the other hand, an expansion vector, which seeks to multiply the points of view considered relevant to the composition of social practices in a territory made active—including, in this process, other ways of life and other territories. This duality works as an analytical tool, but should not be understood in a rigid or exclusive way.

These dynamics not only coexist, but are fundamental to understanding urban climate activism, which operates precisely in these tensions.

The political struggle, in this context, is configured as a dispute around processes of activation and composition with more-than-human agencies—not just between human subjects—and, in this movement, generates new existential territories (Guattari 2011).

2.6 Becoming Sensitive: How to Follow Active Worlds

Climate change, as a problem for human beings, does not exist outside of monitoring networks and the various scientific practices that continually add data and interpretations to this event. To exist socially, they need to be translated into sensitive and observable data (Edwards 2013; Tsing et al. 2024). However, climate change is also present through its translation into the design of new objects, the modification of everyday practices—such as the use of plastic in shopping—the incorporation of new technologies, and transformations in architecture and urban thinking. These practices make the phenomenon sensitive: they affect us not only in terms of values, but also by producing new affordances and material constraints.

Producing oneself in the "machine of the Anthropocene" implies recognising the technical environment not just as a political-economic backdrop, but as an active means—as a *faire-faire*. Subtracting oneself from the machine, interrupting it, avoiding it, sabotaging it: all these verbs are present in both the language and practice of those who become active. In the case of urban activists, these efforts are directed precisely against the environment that formed them—and of which they are, at the same time, heirs. As Hage (2012) notes, this is not just an anti-politics against the system, but an alter-politics: the creation of spaces for experimentation in which it is possible to escape dominant socialities and build new forms of coexistence. The direction of these experiments, however, cannot be assessed without considering the concrete constraints to which they respond.

Climate activism therefore needs to continually generate situations that enable performances of activation—occasions in which everything that is present in everyday experience becomes an opportunity to modify material bonds or to be challenged to do so ("should I keep buying from Amazon?"). The problematising situation destabilises the subject and their ties, requiring adjustments in the ways of living between "chains of dependence" and "relations of interdependence"—or, as Stépanoff suggests, between dense networks and diffuse networks.

While Marres (2015) defines these performances as activated by objects and technologies in the domestic space, my aim is to expand this understanding by articulating these performances—in all their ambiguity—to urban spaces and the overlapping networks that constitute them.

2.7 A City of Many Cities

One Saturday morning (20/02/2020), I decided to follow two simultaneous events in Milan on climate change and the prospects for political, collective and institutional action. The first was an event organised by the Arci circle[4] at which representatives of the Democratic Party—the political force behind Giuseppe Sala's administration—would present the party's line on the energy and ecological transition in the EU. The aim was to present the proposals and challenges of the ecological transition across Europe and how the party's MPs were working to promote the ecological shift in the European Union. The second event was that of a group of students—mostly from the Resilient Gap collective at Milan Polytechnic—residents, members of the Milan Greens (*Verdi*) and onlookers who gathered in the square for an action-event against the destruction of Parco Bassini—by decision of the Polimi and connivance of the city administration.

These two events, although distinct, were linked not only by their simultaneity, but also by a contrast that revealed two divergent ways of responding to climate change, the ecological transition and the role of politics. During the first event, a sentence uttered by a Milanese MP summed up this distinction of positions well: "It was necessary to abandon youthful naivety and propose sensible and responsible responses". The MP recognised the role of young people in pushing for change, but argued that, from that moment on, it would be up to institutional politics to translate the cry from the streets into concrete legislative procedures. A few weeks earlier, the mayor of Milan had declared that

[4] The Italian Recreational and Cultural Association, abbreviated to ARCI, is an Italian social promotion association founded in Florence on 26 May 1957. It generally recognises itself in the ideals of the anti-fascist centre-left.

activists could not just be "environmentalists against" but should also be "environmentalists in favour".

This institutional discourse was contemporaneous with what is known as the "Bassini affair". Between December 2019 and February 2020, several demonstrations were organised, until the Covid-19 health emergency interrupted the mobilisations. However, the Bassini case was not the only focus of activists' attention. Other relevant episodes include the Goccia case, the San Siro stadium and the Baiamonte case—all closely linked to property expansion projects, "green" architecture and, as activist Federica defined it, "the increase in land consumption and the cementification of the city of Milan". In short, these conflicts converged in criticising what the activists called the "Milan Model".

The outbreak of the pandemic has interrupted what promised to be a fruitful articulation between young climate activists and territorial committees engaged in the defence of the so-called green common good. Paradoxically, Milan has stood out, both nationally and internationally, as a model city in green policies: expanding green areas, modernising heating systems, combating air pollution, reducing car traffic, promoting cycling, food security policies, encouraging urban agriculture and advanced waste management. This apparent dissonance, however, should not be read as a contradiction. The same administration that cuts down trees is the one that plants new ones. The residents and activists gathered in Bassini were protesting against this model, trying to get passers-by and neighbours to join the movement. For them, the defence of the park—which, after eight months of mobilisations, would be permanently eradicated to make way for a new physics building at the Politecnico di Milano—represented an attempt to preserve a territory outside the logic of real estate expansion. Although intense, the protests and demonstrations were not enough to prevent its final destruction.

References

Chakrabarty, Dipesh. (2020) When the global reveals the planetary: Bruno Latour interviews Dipesh Chakrabarty. In: Latour, Bruno; Weibel, Peter (eds.). *Critical zones*: the science and politics of landing on Earth. Cambridge and Karlsruhe: MIT/ZKM.

Charbonnier, Pierre. (2020) *Abondance et liberté*: une histoire environnementale des idées politiques. Paris: La Découverte.

De la Cadena, Marisol (2015). *Earth beings*: ecologies of practice across Anden worlds. Durham and London: Duke University Press.

Debaise, Didier. (2017) *Nature as event*: the lure of the possible. Durham and London: Duke University Press.

Debaise, Didier; Stengers, Isabelle. (2021) Résister à l'amincissement du monde. *Multitudes*; 85: 129-137.

Despret, Vinciane. (2004) The Body We Care for: Figures of Anthropo-zoo-genesis. *Body & Society*; 10(2-3):111-134.

Dutreuil, Sébastien. (2020) Gaia is alive. in: Latour, Bruno; Weibel, Peter (eds.). (2020) *Critical zones*: the science and politics of landing on Earth. Cambridge and Karlsruhe: MIT/ZKM.

Edwards, Paul. (2013). *A vast machine*: computer models, climate data, and the politics of global warming. Cambridge: MIT Press.

Gabrys, Jennifer. (2014) A cosmopolitics of energy: diverging materialities and hesitating practices. *Environment and Planning A*, volume 46, pages 2095 – 2109.

Gabrys, Jennifer. (2020) Sensing a moving planet. in: Latour, Bruno; Weibel, Peter (eds.). *Critical zones*: the science and politics of landing on Earth. Cambridge and Karlsruhe: MIT/ZKM.

Gilbert, Scott F. et al. (2012) A symbiotic view of life: We have never been individuals. *The Quarterly Review of Biology;* 87: 325–341.

Grasso, Marco. (2022) *From big oil to big green*: holding the oil industry to account for the climate crisis. Cambridge: MIT Press.

Hage, Ghassan. (2012) Critical anthropological thought and the radical political imaginary today. *Critique of Anthropology*; 32(3): 285-308.

Hawkins, Gay (2011) Packaging water: plastic bottles as market and public devices, *Economy and Society*, 40:4, 534-552.

Kirksey, Eben; Schuetze, Craig; Helmreich, Stefan. (2012) Introduction. in Kirsey, Eben (ed.). *Multispecies salon.* Durham and London: Duke University Press.

Lapoujade, David. (2020b) *William James*: empiricism and pragmatism. Durham and London: Duke University Press.

Latour, Bruno. (2013) *Inquiry into the modes of existence*: an anthropology of the moderns. Cambridge: Harvard University Press.

Latour, Bruno. (2017) *Facing Gaia*: eight lectures on the new climate regime. Cambridge: Polity Press.

Latour, Bruno. (2018) *Down to earth.* politics in the new climate regime. Cambridge: Polity Press.

Latour, Bruno; Weibel, Peter (eds.). (2020) *Critical zones*: the science and politics of landing on Earth. Cambridge and Karlsruhe: MIT/ZKM.

Malm, Andreas. (2016) *Fossil capital*: the rise of steam power and the roots of global warming. London: Verso.

Maniglier, Patrice. (2020) How many Earths? The geological turn in anthropology. *The Otherwise*, n.1, 61–75.
Marres, Noortje. (2015) *Material participation*: technology, the environment and everyday publics. London: Palgrave Macmillan.
Mitchell, Timothy. (2011) *Carbon democracy:* Political power in the age of oil. London: Verso.
Stengers, Isabelle. (2005) Introductory notes on an ecology of practices. *Cultural Studies Review*, 11(1), 183–196.
Stengers, Isabelle. (2011) *Thinking with Whitehead*: a free and wild creation of concepts. Cambridge: Harvard University Press.
Stengers, Isabelle. (2015a) *In catastrophic times*: resisting the coming barbarism. London: Open Humanities Press.
Stengers, Isabelle. (2020b) *Réactiver le sens commun*: lecture de Whitehead en temps de débâcle. Paris: Les empêcheurs de penser en rond.
Stengers, Isabelle. (2021) Putting Problematization to the Test of Our Present. *Theory, Culture & Society*, Vol. 38(2) 71–92.
Stengers, Isabelle. (2020b) We are divided. *e-flux journal*, issue 114, December, 01–06.
Strathern, Marilyn. (1988) The concept of society is theoretically obsolete. In: Ingold, Tim (ed.). *Key debates in anthropology*. London and New York: Routledge.
Tsing, Anna. (2015) *The mushrooms at the end of the world*: on the possibility of life in capitalist ruins. Princeton: Princeton University Press.
Viveiros de Castro, Eduardo. (2019) On Models and Examples Engineers and Bricoleurs in the Anthropocene. *Current Anthropology*; 60:S20: S296-S308
Viveiros de Castro, Eduardo; Danowski, Déborah. (2014) *Há mundo por vir?* Ensaios sobre os medos e os fins. Florianópolis: Cultura e Barbárie/ISA.
Viveiros de Castro, Eduardo. (2014) *Cannibal Metaphysics*. Minneapolis: Univocal Publishing.

CHAPTER 3

Resilience and Climate Urbanism

"The mansion of modern freedoms stands on an ever-expanding base of fossil-fuel use. Most of our freedoms so far have been energy-intensive" (Chakrabarty 2009: 208).
"D'une façon plus générale, les menaces qui pèsent sur la biosphère, la poussée démographique mondiale, la division internationale du travail conduiront les opinions publiques urbaines à penser leurs problèmes particuliers sur fond d'écologie planétaire" (Guattari 2013: 35).
"Whatever the verdict, we have thus to envisage, borrowing Anna Tsing's expression, living in capitalist ruins, that is, in the ruins of the socio-technical infrastructures that ensured our business-as-usual life" (Stengers 2021: 82).

In this chapter, I will follow some performances of activation of non-human agencies and analyse how the explicitness of these agencies contributes to redefining certain local territorial arrangements. I propose different scenarios—or problematisation situations—in which it is possible to trace the emergence of an environment that is both threatened and threatening, as well as the constitution of a specific type of subject called upon to respond to this environment: the climate citizen. This figure emerges at the intersection of moral obligations, technological devices, competences and material resources. Climate citizenship is conceived here as a set of rights and duties that seek to (re)define the relationships

D. Delatin Rodrigues, *Climate Change and Political Action in Italy*,
https://doi.org/10.1007/978-3-031-98771-7_3

considered legitimate between subjects and territories. The chapter is structured as an ethnographic collage of events: official actions promoted by the Municipality of Milan, institutional documents and participatory workshops. The analysis covers the period between 2019 and 2021.

3.1 Becoming a "Climate Citizen"[1]

Historian John Foot ends his book on the city of Milan by expressing his bewilderment at how little attention was paid to the environmental issue by citizens and local authorities: "*Despite the closure of large factories, the pollution situation was dramatic. Massive private motorisation and the failure of local administrations (…) only made the situation worse*" (Foot 2001: 201). He concludes that the city is experiencing a daily ecological-environmental catastrophe (idem: 206).

Since the publication of that book, it can be said that a lot has changed. The city's environmental quality remains unsatisfactory, but it would be unfair to say that both the municipal administration and citizens have remained inert in the face of this reality. In several cities around the world, managers have been adopting policies to mitigate and adapt to climate change (While and Whitehead 2013; UN Habitat 2011; Marvin and Rutherford 2018; Madsen and Hansen 2019), and Milan is part of this movement. Similarly, citizens have not been passive spectators of these processes. An emblematic example is the group *Cittadini per l'aria* (Citizens for the Air), active since 2015, whose actions go far beyond counter-information campaigns about the city's poor air quality.

For those who live in Milan, the name "Cittadini per l'aria" is probably familiar. It's an association that defines itself by a particular form of activism. Observing the Milanese territory—its specificities and problems—and following the way this group intervenes in public issues that affect collective well-being allows us to explore, through a sociological and political lens, new perspectives on what it means to *become active*. And, even more so, what it means to *activate a sensitivity* to common

[1] Our exploratory definition of climate citizens is inspired by the work of Andrea Muehlebach (2012), in which she shows how the rise of voluntarism (sometimes called "citizens with a heart" by its informants) moulds, through different practices and devices (legal, economic and political), a specific form of citizenship that she calls neoliberal.

dimensions of collective life—such as air quality—that are constantly under threat.

John Dewey (2016) proposed a precise definition of the idea of the public: for him, its existence should not be assumed separately from the issues that affect it. The public is not a generic given, but a contingent and situated formation that emerges around specific problems. According to Dewey, the public is constituted when actions—often unregulated—generate consequences that affect a community, or at least part of it. It is in this sense that the actions of *Cittadini per l'aria* can be interpreted: as the emergence of a public around an unresolved problem, which mobilises new ties, sensitivities and forms of engagement.

In other words, the public emerges from a problem whose cause is not directly attributed to specific citizens, but whose effects affect them directly and concretely. Based on a dramatic situation of atmospheric pollution, with devastating impacts on human health in the Milan region, the *Cittadini per l'aria* group constituted itself as a public directly interested in resolving this problem. The harmful effects of pollution are experienced in a localised way—they are linked to a specific territory and affect its population unevenly.[2]

The way a public respond to a problem doesn't necessarily follow a manual of what would be effective or not. This response depends on the characteristics of the public itself, but also on the environment in which it is constituted and the means available to demand solutions at the source of the problem. Faced with the continual production of conditions that are harmful to life, there are many ways to become active and intervene in issues that at first do not seem to concern us. The Cittadini per l'aria association works precisely in this direction: it transforms the abstract generality of urban pollution into concrete and sensitive forms of perception. It makes the phenomenon visible, measurable and experiential. To do this, it inserts technological devices into the daily lives of citizens, turning them into an experimental set—capable of generating relevant data on the effects of air pollution in cities, through the streets, neighbourhoods and their own bodies. This is undoubtedly an action to raise awareness. But perhaps it's more accurate to say that these

[2] https://wsimag.com/it/economia-E-politica/50675-cittadini-per-laria

are practices that transform everyday experiences into opportunities for socio-ecological monitoring.[3]

The subject becomes active as he starts to consider the agencies that make up the urban environment: he learns to inhabit the city, taking into account not only its visible aspects, but also those invisible elements that previously functioned as a neutral and unproblematic backdrop to his daily life. It is a practice of making the ecological and climatic conditions of urban life explicit.

Becoming active in the face of a problematic issue, as exemplified by the actions of the Cittadini per l'aria, means creating technical and material means that enable others to become sensitive to the correlation between air pollution and collective unease. The common good, in this context, is always a creative endeavour. As Isabelle Stengers observes, "there is no common without a doing in common"—it's not just about adhering to an abstract idea, but about building shared practices around something that matters to a community. In other words, it's about forming collectives around something that calls them to think and act together.

"Capacities", as Dewey (2016: 226) reminds us, "are limited by the objects and tools at hand". Recognising this implies understanding that political and ecological activation does not take place in a vacuum, but is mediated by materials, devices and infrastructures that shape both perception and the possibility of action. In the city of Milan, I will analyse the administrative and legal procedures through which the environment and climate increasingly become the subject of public policy, no longer as an externality, but as an active part that interferes in the present and future of the city and its residents and users, culminating in the Aria-Clima Plan (ACP)[4] in 2021—which, due to opposition from centre and

[3] The citizens of the air are not alone. Other groups have also been activated to become capable of monitoring the air by self-constructing air quality control devices, such as OffTopic, a busy social centre that held different workshops during the research period. See: https://www.offtopiclab.org/workshop-autocostruzione-centraline-al-lambrock/

[4] "ostruzionismo messo in campo giovedì 15 luglio dall'opposizione in Consiglio Comunale ha avuto la meglio: 300 emendamenti presentati e la scelta, permessa dal regolamento per un provvedimento all'anno, di non contingentare i tempi di discussione della delibera. The result, considering also the elections for the post and the next limited session of the Council, was to restrict the majority to spending the piano and leave the work for the next consultation". https://www.ecodallecitta.it/milano-salta-lapprovazione-definitiva-del-piano-aria-clima-in-consiglio-comunale/

far-right parties, was not approved. My aim is not to analyse the document, but simply to explore some of its consequences in relation to what it proposes to experiment with and the capacities it presupposes—that will be following by a growing interest from the local media, by the succession of museum exhibitions and artistic events.

This set of practices—political, artistic-cultural, legislative, regulatory and economic—constitutes the conditions through which climate citizenship emerges. For its material, subjective and political fabrication, all available means are considered potentially useful. It's not just a question of tackling energy or pollution problems: food, mobility (with electric cars, scooters and shared bicycles spread across various areas of the city), leisure, urban agriculture and commerce have also been reconfigured. Every aspect of urban life has been touched, and each has been given new definitions and performance capabilities, now assessed according to levels of energy consumption and greenhouse gas emissions.

These initiatives are not isolated actions. There are many "climate citizens" involved. Between the end of 2018 and 2021, I was able to map more than a hundred activities carried out in the city of Milan with an explicit focus on climate change.

It is in this context that interest in the concept of resilience stands out. Originally formulated as an analytical category to measure the carrying capacity of ecological niches (Holling 1973), resilience has become a technical tool for urban planning, the everyday language of various events and disciplines, a political programme and even an ideal of the collective future. From the point of view of this research, resilience is not a concept that explains—rather, it is what needs to be explained. It needs to be investigated based on the other concepts, practices and technologies that it presupposes: guaranteeing conditions for reproduction in the face of extreme events.

During one of the field activities, I had the opportunity to talk to members of the Resilient Cities Office (RCO). I requested this first meeting with the aim of introducing myself, getting to know the activities carried out and exploring the possibility of carrying out part of the ethnographic research in the context of project management. Over the course of more than 24 months following climate initiatives in Milan, I was able to observe the recurring participation of the RCO in events, reports and seminars dedicated to analysing the climatic and ecological risks faced by the city, as well as the plans and challenges associated with urban preparedness for extreme events.[5]

The resilient city must be built piece by piece. This doesn't just involve changing the town planning code or adopting new sustainability indicators but also producing political and social divisions—between those who fully or partially support these measures and those who completely reject them, often because they see them as a justification for state control over private life.[6]

But the divides are not just political or social—they are also physical, inscribed in the urban territories themselves. Using the image proposed by anthropologist Knox of "thinking like the climate", the creation of the Resilient Cities Office (RCO) has enabled the city administration to gather ecological and climatic data based on a logic of governance guided by the effects of climate change, understood as shocks and stresses. Based on this data, future scenarios have been drawn up for the city—if no action is taken—and possible adaptation and mitigation measures have been indicated.

This data, however, also shows that living in the same city can lead to radically different experiences—differences that are not only explained by the supply of services or the proximity of infrastructure. The uneven distribution of heat, hydrological risks, the concentration of pollutants in the air, access to green areas and exposure to flooding have all been mapped—largely by the Politecnico di Milano—transforming the work of urban planners into a literal practice of socio-ecological and atmospheric design (Graham 2018). As an RCO representative said during KlimatFest (2021), "the urban intervention was an experiment that must wait to be evaluated as fair or not".

The very notion of "experience" invites us to reflect on the expectations placed on those who are called upon to live it. Climate urbanism is inseparable from the production of climate citizens: subjects who must

[5] On the website of the Municipality of Milan, we learn that the strategies were "launched in December 2017 as part of the international 100 Resilient cities project (a programme the city joined in 2014 under the Pisapia administration) promoted by the Rockefeller Foundation".

[6] According to the city council of the far-right Fratelli d'Italia party, the administration wants to turn the city into Mao-Tse-Tung's China, where everyone has to commute by bicycle.

respond, participate, adapt—in short, experiment with new forms of personal and collective existence.

A productive analytical key to understanding these processes can be found in the correlated categories of scalability, described by Anna Tsing (2015), and dependency chains, developed by Isabelle Stengers (2020). Both are categories used to read and frame ecological arrangements—abstracting them from their situated contexts and the multiple agents that constitute them. But at the same time, they are also operative devices: they become effective to the extent that they partially capture and transform the environments in which they operate, converting "resources" or "goods" into elements of systemic reproduction.

As Tsing notes, "Making projects scalable takes a lot of work" (2015: 38). Scalability requires simplification: certain operational, material and subjective aspects need to be brought together and emphasised, while others are disregarded or disinvested in. Tsing adds that " the ability to make one's research framework apply to greater scales, without changing the research questions, has become a hallmark of modern knowledge"; and further: " the expectation of scaling up is not limited to science. Progress itself has often been defined by its ability to make projects expand without changing their framing assumptions" (2015: 38).

Here we are facing one of the most intricate dimensions of hegemonic responses to climate change—and this seems to be a central point of climate urbanism. "scalability, in contrast, is the ability of a project to change scales smoothly without any change in project frames (…) scalability banishes meaningful diversity, that is, diversity that might change things" (idem, 38).

What we see, therefore, is that climate urbanism often operates through selective emphasis: relevant aspects of certain agents are amplified, while others are silenced or neutralised. The scalability of a model requires both simplification and disinvestment. However, as the central argument of this analysis, it is important to maintain that the scalability of dependency chains does not eliminate—or neutralise—the existence of other agents capable of destabilising it, be they human or non-human (Tsing et al. 2020).

This is not about resistance in the classic Foucauldian narrative sense, but about the emergence of events, situations and agents that reintroduce contingency and instability. By ignoring what might disrupt its logic, the scalable model does not become immune: it remains vulnerable to unintended consequences and forms of agency that are beyond its control.

When terrestrial agentivity breaks out, scalability can only be structured from tenuous and unstable conditions: it is precisely the creation of territories, where the conditions of habitability (not for everyone and therefore selective) and the circulation of goods and flows of energy and materials can be maintained (Broto and Westman 2019).

3.2 "Living in Milan in Times of Climate Emergency"

On 23/11/2019, I took part in an event entitled "*Vivere a Milano al tempo dell'emergenza climatica*" (Living in Milan during the *climate* emergency) organised by the RCO as part of the "Milano Partecipa" festival, promoted by the Milan City Council and the Cariplo Foundation.[7] The aim of the festival was to "stimulate debate and participation in public life and create awareness of the role that each person can play, to nurture a renewed sense of belonging and community".[8]

The event was organised in the form of a workshop, open to all citizens, so that they could "learn about the work of the 'resilient city' office and discuss measures to combat climate change at a macro and micro level", as one of its officials put it. To this end, they discussed the measures that should and could be carried out by the city administration, as well as individual choices regarding mobility and food. But how exactly do you build a resilient city? Two weeks later, on 13 December 2019, Pellizzaro took part in another event with the same title, where he talked to some representatives of VinoVa, Fridays For Future, Legambiente and Futuro Fogazzaro about climate change and local actions that can be effective.[9]

At the latter event, the city administration's perspective on what could—and could not—be done was made clear. It was stated, for example, that a "zero investment in fossil fuels" would not be the city's responsibility, but a decision to be taken by the Italian state. The resilient city, in this context, reveals a division of responsibilities that,

[7] The Cariplo Foundation plays a key role in supporting climate change initiatives in Milan.

[8] https://www.fondazionecariplo.it/it/news/istituzionali/milano-partecipa-tre-giorni-di-incontri.html.

[9] https://www.youtube.com/watch?v=V-VrDYgjjIg.

although institutionally coherent, tends to prove difficult to sustain in the long term. It is also a way of shifting responsibility to other levels of government.

At the workshop organised by the Resilient City project team in November 2019, relevant clues emerged about how the public administration had been formulating the challenge of simultaneously transforming urban-material flows and the daily practices of the inhabitants. The presentation followed a standardised model: it began with a brief description of the problems faced by the city, went on to take stock of the actions already taken and culminated with a presentation of future plans to adapt the city to extreme weather events—both local and global. The narrative structure followed the familiar script of translating global climate change into the tangibility of local daily life.

The climate phenomenon was explained mainly through quantitative data: graphs, percentages and time comparisons that correlated temperature variations and greenhouse gas emissions. The aim of the workshop was precisely to make this globally abstract data visualisable in everyday life, connecting it to personal and collective practices.

This type of visualisation, however, also frames the possible ways of responding to the climate phenomenon. The multiplicity of ways in which this framework can be constructed acquires a performative character (Callon 2006; Jasanoff 2004), as it defines causal links, delimits critical points and directs attention to certain interpretations of the problem—not revealing a neutral "reality", but a specific reality, mediated by expertise and measuring devices.

This data is isolated, processed, recoded and reassembled, producing a specialised image of the climate phenomenon: an image that inscribes it in everyday life, in economic activities and in the rules that regulate impacts and responsibilities. This is a hegemonic way of presenting the problem, dominated above all by logics from physics and economics, which tends to limit the emergence of alternative narratives—be they other ways of measuring, visualising or experiencing the climate (Urry 2011; Jasanoff 2004; Bonneuil and Fressoz 2017; Tsing et al. 2020).

As Szerszynski and Urry (2010: 04) point out, the territorialisation of climate change is always social: it involves practices, values, scales and situated forms of perception and response:

> "Any description and prediction of climate change and its impacts is entangled with specific imaginaries of how society is, and how it ought to be;

similarly, even the most likely technical of suggested responses will carry with it certain ideas of society. Climate change is always already social; the social does not need to be added to it, just to be revealed".

In addition to the prominent role of natural scientists and physicists, there is a growing participation of urban planners and architects in the public debate on how to organise and design the Anthropocene city (Dawson 2017; Scott 2016; Mostafavi 2016; Turpin 2013). This increased presence makes it possible to identify some relevant clues about the priority actions considered and the technical—and spatial—references through which the work of climate transition, adaptation and mitigation is problematised and translated into concrete interventions.

It is up to urban planners, architects and designers to convert abstract data into tangible solutions, into articulated materialities that can support new practices and forms of urban life.

In the case of Milan, as part of the 100 Resilient Cities network (Galderisi et al. 2020), five critical dimensions of urban climate policy were identified: environmental degradation, extreme heat, shortage of affordable housing, flooding caused by heavy rainfall and civil unrest.[10]

This type of intervention is fundamentally based on spatial transformation, as illustrated by various initiatives underway in the city. On the

[10] Since 2009, the Municipality of Milan has been part of the international network C40 Cities Climate Leadership Group, made up of the mayors of the world's leading cities to promote the process of decarbonisation at urban level. The Municipality joined C40's "Deadline 2020" programme in 2017, pledging to become a carbon-neutral city by 2050 (with the elimination of fossil fuels), to help keep the planet's temperature rise to 1.5 °C (in line with the objectives of the 2015 Paris Agreement). The initiative envisages faster and more incisive action to mitigate climate change, with a plan targeting 2050 and an intermediate target of 2030. (CAP, 9) "SHOCK STRESS—Flooding—Collapse of the infrastructure system—Deprivation and poverty—Phishing attacks—Accidents involving hazardous materials. Stress—Air and water pollution—Heat waves and extreme heat—Degradation of urban spaces—Management of shipbuilding and temporary uses—Social exclusion and inequalities—Lack of decent and affordable housing—Ageing population—Lack of security and possible terrorist attacks—Waves of migration—Riots or civil unrest (757). Based on the identification of these critical aspects, the plan was structured into 6 intervention areas: Cool Milano: urban forestry, agriculture and climate change; Abitare Milano: social innovation, housing, youth and training; Milan City of Water: hydrogeological risk, governance, reopening of Navigli; Circular Milan: urban metabolism, waste, urban and industrial economy; Safer Milan: emergencies, security and cybersecurity; Next Us: infrastructures, requalification and public space, underground services, immigration (757).

one hand, there are large-scale projects that require intensive and extensive interventions—such as the urban regeneration plans for entire areas, exemplified by the Scali ferroviari in Farini and San Cristoforo. In addition to these, there are regulatory instruments linked to public policies in the fields of food, energy and mobility.

On the other hand, there are smaller-scale interventions based on the logic of tactical urbanism (Dawson 2017). These actions seek to generate tangible benefits for residents—and often also influence the property valuation of areas—by altering established local dynamics and allowing new uses for urban space. This is a reconfiguration of the functionalities of urban materiality. A recurring example are measures to close or restrict car traffic, which open up space for alternative forms of use and coexistence in the same territories.

Actions to mitigate and adapt urban infrastructures are always permeated by projections, expectations and desires about the future, interrelated with materials, infrastructures and technologies (Anand et al. 2018; Hetherington 2019). This kind of project cannot be done without considering the simultaneous resistances and possibilities of both subjects and materials (Marvin and Rutherford 2018). According to Pellizzaro, the Milanese leader of resilient cities, "We work on integrating climate risks, bringing first-level knowledge of what is happening in the city. Because in order to plan a city in relation to climate impacts, we need to know how much temperatures have risen and what the rainfall trend is in the urban environment, for example".[11]

The answers provided by these projects reconfigure the environment, establishing new premises and dispositions through which climate citizens begin to consider both their personal dimension ("how do I move around the city?") and their collective dimension ("what impact do my actions have on the lives of others?").

In this context, both climate urbanism and climate activism should be thought of as immersed in unpredictable futures (Braun 2014), in which different imaginaries play a central role (Dobraszczyk 2017). Different political ecologies emerge from the practices and definitions of different actors (Zierderman 2016; Cohen 2017). This political ecology need not, however, be restricted to an anti-political perspective; it can be thought of

[11] https://valori.it/milano-crisi-climatica-urbanismo-tattico/.

as an alter-politics (Hage), centred on experimentation and the invention of alternative forms of collective life.

The aim of the workshop was to present the city of the future—a transformed city, with new ways of living and new habits. The resilience devices presented were seen as the beginning of the formation of new climate citizens, subjects who would be trained in new evaluative capacities, capable of choosing practices and forms of consumption that were less harmful to the environment and the climate.

3.3 Resilience Devices and Dispositions

As we saw in the notes on the workshop, the event was structured as a set of descriptive and prescriptive data, with the aim of helping citizens to become "climate responsible" (RCO official), even considering the diversity of knowledge and profiles present there: self-employed workers, civil servants, students, pensioners. As it turned out, I was the only foreigner in the room. Some of the participants were carrying out practices directly aligned with the resilient city guidelines, such as replacing cars with bicycles.

This is a central point: the need to simplify the discourse by presenting incontrovertible data can limit alternative ways of formulating the problem. As Knox (2020: 47–48) observes:

> "Once the whole number is established, we can begin to divide up responsibilities for carbon emissions reductions" (…) This has the powerful effect of making global climate change amenable to management at a variety of different levels - from the global to the national, local, institutional, and even individual levels".

My intention is not to deny the validity of this approach, as it is part of the scientific method: to focus attention on certain aspects that allow causal chains to be established. But as Alfred Whitehead warns: "concentrate on that observation, dismissing from consciousness all irrelevant modes of experience. But there is no irrelevance" (1968: 74).

The resilient city deals with phenomena that potentially concern everyone, but it cannot control how this relevance will be experienced. As Amin and Thrift (2017: 27) note, "the city can only be partially known"—especially in urban environments sustained by short and long extractive chains. Calculating consumption and emissions is a valid abstraction, but it cannot capture how fossil fuels are experienced, nor the dilemmas, losses and resistance linked to abandoning or insisting on their use. It assumes a homogeneous environment for action, when in reality this environment is unequal. The goal of carbon control drives the design of new infrastructure networks and spatial adaptations to reduce consumption and emissions (Hodson and Marvin 2010; Bulkeley 2013; Blok 2014; Mostafavi 2016). In so-called global cities, the climate transition is increasingly treated as a market opportunity rather than a limit to capital accumulation (Whitehead 2013). However, there is no shortage of criticism of this hegemonic model (Goh 2015, 2020; Dawson 2017; Aylett 2010a; 2010b; Cohen 2015), which points to impasses related to power, resistance, lock-in and experimentalism.

Centred on the mechanisms that should guarantee the transition to a new model of social and ecological reproduction, these criticisms call for a transformation anchored in climate justice—which includes not only environmental policies per se, but also classic issues of urban struggles: the right to housing, transport, social assistance, among others. As Cohen (2017, p. 141) notes, it is "no longer possible (or desirable) to deeply distinguish social from environmental politics".

Outside the socio-technical networks of green enclaves, how are the transition processes experienced by those for whom it represents a critical burden? In parallel with the emergence of extreme climate events, a heterogeneous public is also emerging that problematises these processes from diverse and often experimental perspectives (Gareth and Bulkeley 2017).

The focus on carbon control or the transition to low-carbon societies also invites us to reflect on the analytical tools used to interpret these experiences. Paterson and Mueller (2018), like Cohen (2017), propose an indirect approach to climate policy, analysing its lateral or contradictory effects. This approach makes it possible to highlight tensions in traditional urban policies, as in the case of densification in American cities or housing policy in São Paulo. Similarly, Silver and Marvin (2018) address the low-carbon energy transition in urban Africa as a development problem; and Luque-Ayala (2018), Luque-Ayala et al. (2018a; b) discusses do-it-yourself solar heating systems in Guarulhos (Brazil) from a post-developmentalist perspective.

We can therefore observe conflicts, alliances and the formation of coalitions as central elements in driving socio-technical transitions. This is the case with the Paris municipality's efforts to build an energy policy with decarbonisation targets (Marvin and Rutherford 2018), or the transition process in Greater Manchester (Hodson et al. 2018). There has also been growing research into the circulation of climate urbanism as a global model mediated by transnational networks. Within the literature on global cities, the works of Blok (2014) and Goh (2015) stand out, focusing on the production and diffusion of the "global model" of climate urbanism[12] of urban mitigation and adaptation to climate change.

The relationship between urbanisation, the expansion of the extractive market economy and climate change is intrinsic (Mitchell 2011; Malm 2016; Bonneuil and Fressoz 2017). The period known as the "Great Acceleration"—which began post-1945—marks an unprecedented increase in energy consumption, the use of water resources, land use conversion and deforestation. These processes are directly associated with the expansion of urban infrastructure: dams, transport and telecommunications systems, as well as the intensive use of fertilisers in agriculture—the so-called green revolution (Steffen et al. 2015).

As Hodson and Marvin (2010, p. 300) note, "cities are the material representation of today's energy-intensive economies where carbon-based energy systems—oil, electricity and mobility systems—have made the huge agglomerations of cities and modern industrial systems possible" This reveals that the Anthropocene machine is first and foremost a functional arrangement of networks and infrastructures. As Amin and Thrift (2017, p. 34) state, "one of the main products and producers of Anthropocene".

In this scenario, the "urban" ceases to be just the stage for transformations and becomes the object to be transformed. The Air and Climate Plan[13] was an attempt to establish an integrated plan that should articulate the need to reduce atmospheric pollution with those related to climate mitigation and adaptation policies, "enabling synergistic interventions that produce benefits on both fronts". The Plan aimed to produce

[12] The concept of globalisation is used by Roy (2011, p.10) "to recover and restore the wide range of global strategies that are being staged on an urban scale around the world". In some cases, these urban experiments are closely linked to the aspirations of elites and the creation of world-class cities; in other cases, they are instances of globalisation from below (…) as working bodies circulate in search of survival, sustenance and hope".

[13] All documents in the Plan will be cited as ACP; the number of pages will be taken as indicated in the digital document: https://partecipazione.comune.milano.it/uploads/decidim/attachment/file/333/Delibera_79_del_21-12-2021_Adozione_del_Piano_Aria_e_Clima.pdf.

a "zero emissions" city in an integrated way, also "from an environmental and social justice point of view". The urban development model envisaged in the Plan is aligned with the model proposed for dealing with the economic and social impact of the Covid-19 health emergency.[14]

The notion of resilience adopted by the Milanese municipal administration is based on three main pillars. Its stated aim is to develop "environmental, social and economic resilience in the city by increasing the adaptive capacities of Milan's urban system, strengthening governance and promoting investment in adaptation measures" (ACP 2021: 744). Resilience here is conceived not just as a reaction to external stimuli, but as a proactive capacity:

"Proactive capacity, and not just reactive capacity to an external stimulus, can be considered as a premise for the development of a strategy that materialises in flexible and operational actions, oriented towards implementation, through which the objectives of the Strategy are realised in terms of urban, operational and social resilience" (ACP 2021: 751).

The concept of resilience applied to urban intervention derives directly from the theoretical framework of systemic resilience developed by ecologist C. S. Holling (Meerow et al. 2016). A central reference in ecological studies on resilience, Holling offers a starting point for understanding resilience not only as an ecological property, but also as a socio-spatial and political category—even if, in practice, it is often depoliticised and reconfigured as a post-political governance tool aimed at regulating material and immaterial flows (Swyngedouw 2010).

According to Holling (1973: 17):

[14] "It identifies specific actions aimed at integrating and consolidating pre-existing plans and measures: the Sustainable Energy Action Plan, the Territorial Governance Plan, the Sustainable Urban Mobility Plan, the General Urban Traffic Plan, the Urban Parking Plan, the Territorial Chronological Plan, the Local Food Policy, the Three-Year Public Works Programme, the General Services Plan in the Metro, the Building Regulations, the Regulations for the Use and Protection of Public and Private Green Areas, the road provisions in Area B and Area C. Fulfilling the confirmation of Milan's adherence to the new Covenant of Mayors, which relaunches the challenge of reducing CO2 emissions by 2030, the Plan is also an update of the Sustainable Energy Action Plan approved by City Council Resolution no. 43 of 13 December 2018, which sets a CO2 reduction target of 20% by 2020 with regard to climate change mitigation actions. The Plan includes a time horizon of 2025 for a return to air quality limits and 2030 for the consolidation of a low-emission city model capable of coping with climate change through the implementation of adaptation actions, such as the necessary energy transition to a 'carbon neutral' city by 2050".

> "Resilience determines the persistence of relationships within a system and is a measure of the ability of these systems to absorb changes of state variables driving variables and parameters and still persist. In this definition resilience is the property of the system and persistence or probability of extinction is the result".

The central aspect of this perspective is therefore to identify the conditions in which resilience can be effectively produced. As Holling says, it's about "maintaining flexibility above all else". A population, he argues, responds to environmental changes through a series of physiological, behavioural, ecological and genetic adaptations, restoring its capacity to react to future unpredictable changes (Holling 1973: 18).

From this, I can connect this model of resilience to the notion of dependency chains proposed by Isabelle Stengers. For her, " chains of dependency are fragile and often rife with conflict. They are imperatives, certainly, and demand that we neglect what they define as insignificant" (Stengers 2020: 3). Each individual chain "constructs an uprooted notion of dependence", but, "as soon as the chains get bound to each other, they take on a power that none of them has individually, the power of creating dependencies that take on the appearance of inescapable necessities" (idem).

This point is crucial to avoid interpreting the resilient city as a cohesive and functional system. As will be explored later, both the statements and practices of climate urbanism can (and should) be broken down into different components. The idea of resilience attempts to articulate these parts as if they formed an integrated system; however, this is, at best, a partial cut-out of elements that, territorially, can be relevant both for mitigating impacts and for identifying new vulnerabilities.

As Easterling states (2021: ix), "not morphology alone, but the interaction between components, determines the richness of this loose and changeable assembly of part". Resilience, in this sense, is not an intrinsic quality, but an effect to be produced through urban socio-technical transformations. Holling reinforces this perspective by stating that:

"The resilience framework can accommodate this shift of perspective, for it does not require a precise capacity to predict the future, but only a qualitative capacity to devise systems that can absorb and accommodate future events in whatever unexpected form they may take" (Holling 1973: 21).

Although I had approached the members of the project management team directly to check their willingness to collaborate with my research, I was not given this opportunity.[15]

It was not possible, during this research, to investigate how the data made available by the Resilient Cities Office (RCO) was produced internally, nor how the team members liaised with the rest of the municipal administration. In one of the few conversations I had with two of its members, one of them said: "We encountered a lot of resistance within the administration to the idea of making resilience a transversal operator in all sectors of public management" (personal communication, December 2019). On the other hand, when asked about the protests in defence of green areas, mainly motivated by the Bassini case (see Chapter 4), one of the RCO representatives replied that the protesters simply "hadn't understood the municipality's plans", adding that "in the end, the municipality will plant more trees than there were in Bassini".

The Air-Climate plan articulates practices, issues and materialities around the notion of "increasing resilience and promoting the capacity to resist emergency situations" (ACP 2021: 97). The resilience model, as the operative matrix of heterogeneous chains of dependency, aims to contribute materially and discursively to the formation of subjects capable of "questioning themselves and learning to be flexible and open to innovation"; as the programme document states: "adaptation measures require a certain change in our daily habits: knowing how to give up individual practices so that everyone, collectively, has a better chance of living well. It is precisely in the realm of everyday life and domestic practices that social resilience can find fertile ground" (ACP 2021: 767).

This means that in the face of a virus, a storm or a heatwave, both individuals, companies and territories must be prepared to guarantee their continuity. The individual therefore needs to be adjusted to the resilient machine. But as a set of multiple parts, this machine also generates dispositions, understood as "a agency or potential immanent in an arrangement - a property or propensity within a context or relationship" (Easterling 2021: ix).

The individual response, in this sense, is always moulded by the devices and the spatial-relational configuration in which it is inserted. When the means of reproduction are organised by local and global networks whose

[15] The officials suggested that I write directly to Pellizzaro to ask for authorisation to continue the RCO's work, but the email I sent never received a reply.

control is beyond the capacity of individuals to intervene (as shown by Mitchell 2011), the modes of interaction are channelled into forms of adherence, conformity or moralised evasion.

Faced with risks to human life, biodiversity and productive structures, the resilience model can be interpreted as a new technology for governing the precarious (Fassin 2012) or as a model of environmental governance (Luke 1995; Scott 2016), whose aim is to regulate the relationship between climate, population, environment and territory. The ability to provide ecological services (such as clean air or drinking water), the maintenance of which goes beyond the administrative boundaries of the city, requires a redefinition of the very concept of the environment and an expanded understanding of ecosystem support, both on the part of the authorities and activists.

Hodson and Marvin (2009: 194) call this new technical-political arrangement Urban Ecological Security (UES), a concept that designates techniques aimed at "safeguard flows of resources, infrastructure and services (...) informing strategies to reconfigure cities and their infrastructures in ways that help to secure their ecological and material reproduction". Cities are promoted as key spaces for mitigation and adaptation, and the protection of physical and digital urban infrastructures in the face of extreme weather events becomes a priority (Hodson and Marvin 2009, 2010; While and Whitehead 2013; Luque-Ayala et al. 2018a, b).

According to Long and Rice (2018), these transformations are generating a broad change in urban planning paradigms, now orientated towards building resilient and climate-friendly cities. This change makes up what the authors call climate urbanism. A growing body of literature has investigated these new ways of governing climate change in urban contexts, especially in relation to spatial and technological transformations.

This is an aspect that will gain ethnographic density over the course of the book, as I consider the becoming active of subjects who try to respond to climate change from within an environment shaped by the extractive urban machine. The simple act of eating a hamburger raises questions about the origin of the meat, labour exploitation, violence against animals, the simplification of ecosystems and the resources mobilised for its production. In the workshop analysed, food was mobilised as evidence of how "individual consumption practices contribute to emissions" and to the continuity of the socio-economic model that sustains the climate crisis.

It is not a question of denying this causal relationship, but of recognising that the exclusive focus on individual practices makes other dimensions invisible, such as the logistics of the international market or the extractivist model of the global economy. The symbolic and material consequences of these different narratives are profound. Presenting climate change through inversion strategies can be productive: it allows us to make visible what is invisible to some (city dwellers in the Global North), but explicitly experienced by others (as in the case of indigenous struggles against cattle ranching in northern Brazil).

The diversification of the extractive urban machine is the ground on which multiple publics emerge, which will be analysed here as partially urban experiences—a partiality given by the concrete practices of daily reproduction in a relational space structured by extractive infrastructures (Banoub et al. 2020).

What had been called climate urbanism is the configuration of devices and the generation of provisions aimed at managing the heterogeneous flows that cross urban space. And the devices are many: ecological certification of buildings, intelligent lampposts, car, bicycle and scooter sharing via apps; restricting the entry of polluting products, the use of biodegradable bags, green walls and roofs, reusable bottles, "zero waste" shops (25 identified); subsidies for housing renovation, emission control zones, cycle paths, low-consumption household appliances, intelligent thermal control devices, among many others.[16]

Resilience mechanisms must "bring out conscious and resilient individuals" (ACP: 5). The need to promote material and subjective transformations in cities based on the injunction that "the climate changes and the city is obliged to change with it" (ACP: 761). But, as Knox notes with regard to similar plans carried out in the city of Manchester, "climate change risked fundamentally unsettling methods of contemporary governance that administrators were familiar with—methods that built on imaginaries of the human population, markets, and economies" (2020: 18).

[16] Here we can see some of the technologies presented by Simon Marvin: https://www.youtube.com/watch?v=qpwzIgt81hU&t=2547s.

Resilience crosses bodies, but also the capacity for business innovation; it refers both to the soil's ability to absorb impacts and to the subjective dimension of getting up after a shock.[17]

Perhaps we can also speculate that the contemporary appeal of the category of resilience lies partly in its ability to fill a gap in urban governance instruments. Resilience has become an integral element of the "growth machine", working in an abstract and technocratic way: the calculation of resources is often carried out in such a way as to exclude everything that could represent value for the inhabitants—or that could introduce noise into the scalability of the models. In these terms, resilience can be completely removed from the lived and concrete city, as demonstrated by the insistence of activist protests that say: "we need to transform the current development models".

The emergence of multiple problematic situations associated with innovations implemented under the banner of sustainability is therefore foreseeable. These innovations will become the subject of widespread concern and collective debate: the introduction of scooters and shared electric cars; the replacement of energy systems in low-income housing; the social costs of the energy transition for vulnerable populations. The transformation of the urban socio-material fabric will produce unequal effects, generating concerns and mobilisations.

Given the multiplication of these problematic situations, it seems pertinent to follow how they give rise to different publics—momentary collectivities that organise around specific aspects of the anthropocentric urban machine. As Blok and Farías (2016: 06) observe:

"Democratic publics are not the expression of some pre-existing community, such as a nation or a city, but groups of strangers who share only one thing, being affected or concerned by the indirect consequences of others' actions. Such consequences, and the problematic and controversial objects, things and issues in whose shape they materialize, thus enable public life".

It is essential to recognise that these transformations are not just the result of urban planning or institutional policies. The ecological and climatic rethinking of social practices is beginning to contaminate all areas of everyday life. It manifests itself not just discursively, but as a

[17] "Of the rest of the traumatic and negative events, it can always be said: it is the proactivity associated with the process of building adaptive resilience" (ACP: 894).

physical-material presence, through the introduction of devices into social space—and simultaneously being reconfigured by it.

Companies, associations, public organisations, design practices, artistic performances, advertising and political parties—it would be difficult to find a social domain today that has not been affected in some way by this climatic inflection. This is a central aspect of the many contemporary versions of the resilient city.[18] (Despret 2002) of transition practices.

These versions don't just correspond to different visions, but to perspectives that emerge from situated articulations with the environment to which they are linked. Their analytical value lies in highlighting, as Stengers (2020: 27) proposes, "la signification de termes (...) des situations concrètes" "car on n'abstrait jamais en générale, toujours sur un mode déterminé ".

Returning to the development of resilient devices, as outlined in the municipal administration's proposal, we see that they are organised around three pillars, each with its own specificities, as they seek to rationalise the diversity of agentivities involved, transforming them into functionally operable elements.

Environmental resilience is based on the need for design practices to address the multiple climatic and ecological challenges—rising temperatures, flooding, pollution atmospheric—by returning significant parts of the urban territory to vegetation and permeable surfaces, as well as revaluing nature as an element of well-being and biodiversity in the urban context (ACP 2021: 760).

Social resilience, on the other hand, is defined as "the capacity of an individual to generate biological, psychological and social factors that allow them to resist, adapt and strengthen themselves in the face of a risky situation, generating an individual, social and moral outcome" (ACP 2021: 762).

Finally, economic resilience is understood as:

"the capacity of an economic system to return to its initial state after suffering a shock or disturbance. It can also be defined as the condition in

[18] "According to Latour, a theory or a device can be defined as 'occasions' for a version or, to use his terms, they constitute an 'offer of opportunity' directed at a phenomenon (...) a device can be considered as an offer of opportunity for phenomena that, in other devices, would have no chance of existing (...) there will no longer be a body left to itself that is asked to react, but a different scene that is full of 'offers of opportunity'" (Despret 2002: 23).

which the system contains, within itself, adaptive capacities that allow it to spontaneously reorganise itself, after the shock, in its economic, institutional and social structure, and find new avenues for growth" (ACP 2021: 772).

The introduction of devices and new provisions in the urban environment must therefore contribute to "promoting survivability in emergency situations and reducing repair costs" (ACP 2021: 262). Among the critical aspects recognised are the serious social effects caused by disruptive events—the Covid-19 pandemic being a paradigmatic example. In this context, it becomes clear that "an urban system cannot be considered disconnected from the community that inhabits it and, therefore, even with regard to adaptation to climate change, it is essential to contaminate communities with the paradigm of resilience" (ACP 2021: 763).

This "contamination" operates through the multiplication of devices through which climate responses can be materialised. "Climate citizenship" cannot be achieved without the right means for its activation. These means include a variety of institutional and technical devices designed to support individual and collective practices. On the other hand, we also see grassroots initiatives that activate smaller devices, such as permaculture techniques, guerrilla gardening, the creation of common spaces, agroforestry or collective actions like the Bike Strike—a demonstration in which cyclists used the new cycle lane infrastructure to symbolically occupy the central road network, temporarily blocking traffic.

These smaller devices can be situated in what Schlosberg and Coles (2015) call the "new environmentalism of everyday life", characterised by a " sustainable materialist focus on collective practices and institutions of provision of the basic needs of everyday life" (2015: 166). The emphasis on materialist aspects is central, as it builds a bridge between the importance of the socio-technical dimensions of climate urbanism and everyday experience, which becomes the focus—and concern—of these movements. It's not just about behavioural choices or consumption, but about the production, circulation and reproduction of material flows through which these collectives confront problems of power and inequality.

The attention that these movements pay to the flows of everyday life reveals a new materialist politics (Schlosberg and Coles 2015), which can shed light on practices of embodied subjectivity and proposals for social transformation linked to a low-carbon transition perspective—a point that

I believe has been neglected in some of the recent literature on the subject.[19]

Some of these "experiments" can be illustrated by the example of transition towers. Barry (2012: 115) argues that the Transition Towns movement exemplifies a "creative adaptive management" approach aimed at building more resilient communities. These initiatives are based on the fundamental belief that communities can actively shape the socio-ecological and social conditions for their own flourishing.

For North and Longhurst (2012: 1424), transition initiatives operate on a local scale—whether at the level of cities, villages or neighbourhoods—developing community strategies to reduce dependence on non-renewable energies and cut greenhouse gas emissions, while at the same time building low-carbon livelihoods anchored in localised economies. In both approaches, the aim is to modulate socio-technical and spatial transformations through the incorporation (Schlosberg and Coles 2015) of low-carbon practices and purposes into everyday life and at the local level.

However, the problematisation of social change on this everyday-local level raises critical questions related to scale. In the case of certain transition experiences (North and Longhurst 2012) or adaptation projects (Goh 2015; Braun and Wakefield 2014), actions can be considered punctual in terms of scale and ambiguous in terms of their results. Several authors have warned of the risks that such interventions could generate

[19] Evaluating these political actions requires a conceptual distinction about the agents, objectives and means that are mobilised to carry out their actions: in this case, we can see that some scholars have made a very important distinction about so-called *green*, *grey* and *luxury* ecologies. *Grey* ecologies are based on interventions such as infrastructure and high-tech strategies of the built environment that have produced little pollution; *green* ecologies are defined as more traditional policies such as green areas, parks, urban forests that could improve the quality of air, water and well-being (Cohen 2017; Wachsmuth and Angelo 2018); *luxury* ecologies denote projects and developments aimed primarily at benefiting the "professional class and associated businesses, especially finance, real estate and the companies that assist and surround them" (Cohen 2017: 143). These academics are trying to forge a framework that can give density to the collective experience of aiming for a low-carbon transition that does not end up producing more inequalities and vulnerabilities for people without power. Based on ethical and normative questions of equal rights, an idea of democratic ecologies that emerges, in our view, needs to go beyond the human-only approach of traditional political and social science to more explicitly include the position of non-humans in these new urban assemblies—or, as noted by Centemeri (2018: 299), as "multi-species commons" that can be taken as a speculative proposal about possible trajectories that diverge from the anthropocentric model.

new inequalities (Davis 2010), socio-spatial exclusions (green gentrification) and disconnected sustainable enclaves, made possible by urban bypass mechanisms (Graham and Marvin 2001; Hodson and Marvin 2010; Cohen 2017; Hodson et al. 2018).

Actions aimed at reducing emissions, capturing CO_2 (as proposed by urban forests), creating new risk management techniques (e.g. digital sensor systems) or promoting the so-called circular economy must be understood as situated interventions that operate in socio-spatial fields densely made up of multiple actors, interests, coalitions and hegemonic forces. These fields largely determine the contours of socio-ecological urban futures (Zeiderman 2016; Cohen 2017; Wachsmuth and Angelo 2018). As Blok and Farías observe, "the kinds of struggles made visible are those that result from the irruption of ontological and epistemic uncertainties, whose resolution involves the invention of new political arrangements" (2016: 7).

The aim of this chapter was to explore the resilient city from the perspective of the devices and dispositions it seeks to produce. There is no resilient city without resilient citizenship. This aspect has often been neglected in urban studies, which tend to favour public policies and macroeconomic aspects—with important exceptions, of course. I don't suggest that these dimensions are irrelevant, but I do argue that it is necessary to turn our attention to the personal and collective practices that these devices and dispositions generate, and the ways in which they reconfigure the political agency of subjects in everyday life.

References

Amin, Ash and Thrift, Nigel. (2017) *Seeing like a city*. Cambridge: Polity Press.

Anand, Nikhil; Appel, Hannah; Gupta, Akhil. (2018) Introduction: Temporality, Politics, and the Promise of Infrastructure. In: Anand, Nikhil; Appel, Hannah; Gupta, Akhil (eds.). *The promise of infrastructure*. Durham and London: Duke University Press.

Banoub, Daniel, and Sarah J. Martin. (2020) "Storing Value: The Infrastructural Ecologies of Commodity Storage." *Environment and Planning D: Society and Space* 38(6): 1101–1119. https://doi.org/10.1177/0263775820911003

Barry, John. (2012) The Politics of Actually Existing Unsustainability. Oxford: Oxford University Press.

Blok, Anders. (2014) Worlding cities through their climate projects?, *City*, 18(3), pp. 269-286.

Blok, Anders; Farias, Ignacio (eds.). (2016) *Urban cosmopolitics*: agencements, assemblies, atmospheres. London and New York: Routledge.
Bonneuil, Christophe and Fressoz, Jean-Baptiste. (2017) *The shock of the anthropocene*: the Earth, history and us. London: Verso.
Braun, Bruce. (2014) A new urban dispositif? Governing life in an age of climate change. *Environment and Planning D: Society and Space*, vol. 32, pages 49 – 64
Braun, Bruce and Wakefield, Stephanie. (2014) Inhabiting the post-apocalyptic city. *Society & Space*, pp. 01–11
Broto, Vanessa Castán; Westman LK. Ten years after Copenhagen: Reimagining climate change governance in urban areas. *WIREs Clim Change*. 2020;11:e643.
Bulkeley, Harriet, and Vanesa Castán Broto. (2013) "Government by Experiment? Global Cities and the Governing of Climate Change." *Transactions of the Institute of British Geographers* 38(3): 361–375. https://doi.org/10.1111/j.1475-5661.2012.00535.x
Callon, M. What does it mean to say that economics is performative?, *CSI Working papers series* (5), Paris, 2006.
Chakrabarty, Dipesh. (2009) "The Climate of History: Four Theses." *Critical Inquiry* 35(2): 197–222.
Cohen, Daniel Aldana. (2017) The other low-carbon protagonists: poor people's movements and climate politics in São Paulo. In: Greenberg, Mirian and Lewis, Penny (eds.). *The city is the factory*: new solidarities and spatial strategies in an urban age. Cornell: Cornell University Press.
Davis, Mike. (2010) "Who Will Build the Ark?" New Left Review 61 (January–February): 29–46.
Dawson Ashley. (2017) *Extreme Cities*: The Peril and Promise of Urban Life in the Age of Climate Change. New York: Verso.
Despret, Vinciane. (2002) Le emozioni: etnopsicologia dell'autenticità. Milano: Elèuthera.
Dewey, John. (2016) The public and its problem: an essay in political economy. Ohio: Ohio University Press.
Dobraszczyk, Paul. (2017) Sunken cities: climate change, urban futures and the imagination of submergence. *IJURR* Vol. 41 (6), pp. 868-887.
Easterling, Keller. (2021) *Medium design*: knowing how to work on the world. London: Verso.
Fassin, Didier. (2012) Humanitarian Reason: A Moral History of the Present. Berkeley: University of California Press.
Foot, John. (2001) Milan Since the Miracle: City, Culture and Identity. Oxford: Berg.

Galderisi, A., Limongi, G. & Salata, KD. (2020) Strengths and weaknesses of the 100 Resilient Cities Initiative in Southern Europe: Rome and Athens' experiences. *City Territ Archit* 7, 16.

Gareth, Edwards and Bulkeley, Harriet. (2017) Heterotopia and the urban politics of climate change experimentation. *Environment and Planning D: Society and Space* 0(0) 1–20.

Goh, Kian. (2015) *A Political Ecology of Design*: Contested Visions of Urban Climate Change Adaptation. (Thesis) Urban and Environmental Planning, Yale University.

Goh, Kian. (2020) Flows in formation: The global-urban networks of climate change adaptation. *Urban Studies*, Vol. 57(11): 2222–2240.

Graham, S. and Marvin, S. (2001) *Splintering urbanism*: networked infrastructures, technological mobilities and the urban condition. London: Routledge.

Graham, Stephen. (2018) *Vertical*: the cities from satellites to bunkers. London: Verso.

Guattari, Félix. (2013) Qu'est-ce que l'écosophie? Textes (1985–1992), edited by Stéphane Nadaud. Paris: Éditions Lignes.

Hetherington, Kregg. (2019) Introduction: keywords of the Anthropocene. In: Hetherington, Kregg (ed.). *Infraestructures, environment, and life in the Anthropocene*. Durham: Duke Press University.

Hodson, Mike, and Simon Marvin. 2010. "Urbanism in the Anthropocene: Ecological Urbanism or Premium Ecological Enclave?." *City* 14 (3): 298–313.

Hodson, Mike and Marvin, Simon. (2009) Urban ecological security: a new urban paradigm?, *International Journal of Urban and Regional Research*, 33, pp. 193–215.

Hodson, Mike, Marvin, Simon and McMeekin, Andy. (2018) The amenable city-region: the symbolic rise and the relative decline of Greater Manchester's low carbon commitments, 2006–17. In: Luque-Ayala, Andrés; Marvin, Simon and Bulkeley, Harriet eds. *Rethinking urban transitions*: politics in the low carbon city. New York: Routledge.

Holling, C.S. (1973) Resilience and Stability of Ecological Systems; *Annual Review of Ecology and Systematics* 1973 4:1, 1-23.

Jasanoff, Sheila. (2004) Ordering knowledge, ordering society, in: Jasanoff, Sheila. *States of knowledge*: the co-production of science and social order. London and New York: Routledge.

Jasanoff Sheila. (2010) "A New Climate for Society," *Theory, Culture & Society*, 27(2-3): 233-253.

Knox, Hannah. (2020) *Thinking like a climate*: governing a city in times of environmental change. Durham and London: Duke University Press.

Long, Joshua and Rice, Jennifer L. (2018) From sustainable urbanism to climate urbanism. *Urban Studies* 00 (0), pp. 1-17

Luke, Timothy. (1995) On environmentality: geo-power and eco-knowledge in the discourses of contemporary environmentalism. *Cultural Critique*, n.31, p.57–81.
Luque-Ayala, Andrés. (2018) Post-development carbon. In: Luque-Ayala, Andrés; Marvin, Simon and Bulkeley, Harriet eds. *Rethinking urban transitions*: politics in the low carbon city. New York: Routledge.
Luque-Ayala, Andrés; Marvin, Simon and Bulkeley, Harriet. (2018a) Introduction. In: Luque-Ayala, Andrés; Marvin, Simon and Bulkeley, Harriet eds. *Rethinking urban transitions*: politics in the low carbon city. New York: Routledge
Luque-Ayala, Andrés; Marvin, Simon and Bulkeley, Harriet. (2018b) Rethinking urban transitions: an analytical framework. In: Luque-Ayala, Andrés; Marvin, Simon and Bulkeley, Harriet eds. *Rethinking urban transitions*: politics in the low carbon city. New York: Routledge.
Madsen, S. H. J., & Hansen, T. (2019). Cities and climate change—Examining advantages and challenges of urban climate change experiments. *European Planning Studies*, 27(2), 282–299.
Malm, Andreas. (2016) Fossil Capital: The Rise of Steam-Power and the Roots of Global Warming. London: Verso Books.
Marvin, S., & Rutherford, J. (2018). Controlled environments: An urban research agenda on microclimatic enclosure. *Urban Studies*, 55(6), 1143–1162. https://doi.org/10.1177/0042098018758465
Mitchell, Timothy. (2011) *Carbon democracy:* Political power in the age of oil. London: Verso.
Meerow, Sara; Newell, Joshua P.; Stults, Melissa. (2016) Defining urban resilience: A review. *Landscape and Urban Planning* 147, 38–49.
Mostafavi, Mohsen; Doherty, Gareth (eds.). (2016). *Ecological urbanism.* Cambridge: Harvard University Graduate School of Design.
North, Peter and Longhurst, Noel. (2012) Grassroots localisation? The scalar potential of and limits of the 'transition' approach to climate change and resource constraint. *Urban Studies* 50 (7), pp. 1423-1438.
Scott, Felicity (2016). *Outlaw territories*: environments of insecurity/ architectures of counterinsurgency. New York: Zone Books.
Stengers, Isabelle. (2020a) *Réactiver le sens commun*: lecture de Whitehead en temps de débâcle. Paris: Les empêcheurs de penser en rond.
Stengers, Isabelle. (2021) "Putting Problematization to the Test of Our Present." *Theory, Culture & Society* 38(2): 71–92.
Steffen, Will; Katherine Richardson; Johan Rockström; et al. (2015) "Planetary Boundaries: Guiding Human Development on a Changing Planet." *Science* 347(6223): 1259855.
Swyngedouw, Eric. (2010) Apocalypse forever? Post political populism and the spectre of climate change. *Theory, culture and Society*; vol.27(2–3): 213–232.

Szerszynski, Bronislaw and John Urry. (2010) "Changing Climates: Introduction," *Theory, Culture & Society*, 27(2-3): 1-8.

Tsing, Anna. (2015) *The mushrooms at the end of the world*: on the possibility of life in capitalist ruins. Princeton: Princeton University Press.

Tsing, Anna Lowenhaupt; Jennifer Deger; Alder Keleman Saxena and Feifei Zhou. (2020) Feral Atlas: The More-than-Human Anthropocene. Stanford: Stanford University Press.

Turpin, Etienne. (2013) "Introduction: Who Does the Earth Think It Is, Now?" In Architecture in the Anthropocene: Encounters Among Design, Deep Time, Science and Philosophy, edited by Etienne Turpin, 7–42. Ann Arbor, MI: Open Humanities Press.

United Nations Human Settlements Programme (UN-Habitat). (2011) State of the World's Cities 2010/2011: Bridging the Urban Divide. Nairobi: UN-Habitat.

Urry, John. (2011) *Climate Change and Society*. Malden: Polity Press.

Wachsmuth, David and Angelo, Hillary. (2018) Green and gray: new ideologies of nature in urban sustainability policy, *Annals of the American Association of Geographers*, 108:4, pp. 1038-1056.

While, Aidan; Whitehead, Mark. (2013) Cities, Urbanisation and Climate Change. *Urban studies* 50(7) pp.1325–1331, may.

Whitehead, Mark. (2013) Neoliberal urban environmentalism and the adaptive city: towards a critical urban theory and climate change. *Urban Studies* 50(7) 1348–1367.

Zeiderman, Austin. (2016) "Adaptive Publics: Building Climate Constituencies in Bogotá." *Public Culture* 28(279): 389–413. https://doi.org/10.1215/08992363-3427499

CHAPTER 4

The Urban Forest: Green Infrastructure in the Resilient City

4.1 Introduction

> "All the architecture we know is Holocene architecture. Architecture has had to deal with many unpredictable factors, but the Holocene climate has always been an assumed constant."[1] (Wark 2017)
> "Infrastructure itself is a political" (Von Schnitzler 2016: 129)

As we saw in the previous chapter, it has become a widely shared imperative to make explicit the climatic and ecological conditions that sustain life, so that these conditions can be the subject of planning and intervention. In this chapter, I will focus on how the urban forest is mobilised as a central device in the response to the contemporary climate and ecological crisis—and, more specifically, how it is translated as a technology for "green living", including in private contexts such as private homes.

The data presented was collected between 2019 and 2021, based on official public events at which the city of Milan's urban forestry plan was presented, as well as institutional documents, news reports and artistic, architectural and design events at which trees and vegetation were discussed as part of urban planning.

[1] https://www.e-flux.com/architecture/accumulation/122201/from-architecture-to-kainotecture/.

D. Delatin Rodrigues, *Climate Change and Political Action in Italy*,
https://doi.org/10.1007/978-3-031-98771-7_4

My aim is to explore the forest as a device that must be installed in the urban machine to mitigate the adverse effects that this same machine produces at the local level. This analysis will be done by "empiricising concepts, [to] examine how weighty problems of knowledge, politics, and the real acquire their significance in specific settings" (Marres 2015: 160).

4.2 What Forest(s)?

As many analysts have observed, the role of forests in cities is no longer restricted to activist practices of regen2019eration or the creation of community gardens. They are now an important urban policy tool in times of climate change (Blok 2014). The proliferation of vegetation has become a characteristic element of contemporary responses to urban challenges, and different forms of urbanism are disputing its meaning and function (Hodson & Marvin 2010).

Urban nature and green infrastructure have come to occupy a central role in urban planning and environmental studies. As cities face climate change, biodiversity loss and socio-environmental injustices, scholars have critically analysed how urban ecological practices shape power, governance and interspecies relations.

Gabrys (2020) discusses, for example, the digitalisation of urban nature through smart infrastructures. While such technologies improve environmental monitoring and sustainability, they also run the risk of intensifying extractivist logics and social inequalities. The so-called Internet of Nature transforms green infrastructures into digital management sites, raising concerns about the commodification of ecological systems and the algorithmic governance of green spaces. The growing interpenetration between digital and ecological infrastructures has raised questions about whether smart urban environments promote sustainability or merely reinforce existing socio-economic disparities.

In parallel, other researchers have drawn attention to ruderal ecologies, which arise spontaneously in disturbed urban environments. Stoetzer (2018) argues that these ecologies challenge planning paradigms by emphasising the agency of non-human life and the adaptive capacities of marginalised urban communities. The case of Berlin's Gecekondu—an improvised green space on contested land—exemplifies how ruderal formations reconfigure urban governance and propose alternative forms of urban ecology. Unlike planned green spaces, these spontaneous ecologies emphasise non-human resilience and improvisation, demonstrating how nature and culture co-evolve in the face of urban disturbance.

In a different key, Myers (2019) proposes the concept of the Planthroposcene as an alternative to the dominant narrative of the Anthropocene. Instead of considering plants as passive elements in the human-designed urban environment, Myers emphasises the entanglement between plant life and human life in the formation of urban ecologies. Gardens and green infrastructure, she argues, are not just aesthetic or functional spaces, but places where species meet, reflecting wider ecological and political transformations. Planthroposcene proposes a relational reconfiguration of urban nature that recognises plants as agents participating in public life.

The social and economic dimensions of urban greening have also been critically examined. Angelo (2019a, b), for example, questions the assumption that greening is intrinsically positive, arguing that these initiatives often contribute to gentrification and exclusionary urbanism. The commodification of green spaces, she argues, aligns with neoliberal urban policies that prioritise economic growth over social justice. Green infrastructure projects tend to increase property values and displace low-income populations, exposing the paradox of sustainability in contexts of capitalist urbanisation.

As cities continue to integrate ecological and technological solutions into their restructuring plans, a critical approach to urban nature becomes essential to imagining fairer and more inclusive urban futures. As Mostafavi (2016: 17) stated: " we need to view the fragility of the planet and its resources as an opportunity for speculative design innovations rather than as a form of technical legitimation for promoting conventional solutions".

Inspired by this call to reinvent cities, different professionals began to experiment with visions of the future. In this spirit,

> "The municipality accepts the challenge: it will be one of the "Reinventing Cities" taking part in the call for proposals launched by the C40 network, the network of leading cities in the world, calling on architects, designers, entrepreneurs, associations and artists to regenerate five areas of its territory from an environmental and urban point of view".[2]

[2] https://milano.corriere.it/notizie/cronaca/17_novembre_17/milano-5-quartieri-reinventare-parte-bando-internazionale-8f73cc78-cb03-11e7-bd3e-51a6bf213dd1.shtml.

The response to climate change has been shaped around the notion of an urgent challenge: the imperative to transform cities. At the end of 2018, the international exhibition Broken Nature: Design Takes on Human Survival opened at the Triennale di Milano, curated by Stefano Boeri—a central figure in the consolidation of this urban-climate model. The exhibition brought together a wide range of works that sought to give form to the multiplicity of experiences and imaginings linked to the contemporary environmental crisis.

That was precisely the point: climate change needed to take shape, to be named, to become an object of public attention. It was necessary to translate the planetary event into speculative but tangible forms—ones that could be perceived, experienced and disputed. Climate change is thus embodied between collapse and the projection of possible futures. Sensitive and aesthetic beings are generated to affect the environment and make it a shared concern; at the same time, they function as invitations to speculation (individual and collective), as alternatives to the dominant fossil and modernist narrative.

As the exhibition's curatorial manifesto stated:

> "Under these circumstances, it is not enough anymore for designers to be politically and chemically correct. "Organic", "green", "environmental" and "sustainable" are buzzwords that have been applied in earnest to design—including food and fashion—over the last two decades, as have the terms "ethical" and "aware." Yet, despite these noble intentions towards humans, animals, plants and places, we are still—as individuals and communities—tracking a course of destruction through overconsumption and disregard for countless forms of life, including our own. Designers can change flashy trends into more meaningful strategies imbued with agency by encouraging new behaviours using objects—from physical to digital items, speculative scenarios to functional tools, architectures to video games and construction processes for new biological entities to old-school mending kits—as prompts and foils. Citizens of the world have as much responsibility as designers. Broken Nature thus demands us all to collectively apply our resources in order to constructively repair the damaged ecosystems that we inhabit and shape".[3]

The curatorial proposal therefore positions design—and more broadly, material and aesthetic culture—as a crucial field in the translation of the

[3] https://www.brokennature.org/exhibition/theme/.

ecological crisis. It challenges both technical neutrality and emotional passivity by proposing that new ways of imagining and inhabiting the world be catalysed by speculative, sensitive and collective practices. The urban forest, in this context, can be read not only as ecological infrastructure, but also as a political and cultural artefact of an era in search of material reconfigurations of the bond between humans and the planet.

As we will see throughout this chapter, the way in which this new relationship with non-human agency is being constructed does not necessarily imply a break with the extractivist logic that preceded it—since it continues to depend on material and energy resources extracted from outside the urban territory. The forest was designed for life in the Anthropocene (Myers 2019), not against it. In other words, the urban forest doesn't break with the chains of dependency that structure the contemporary city—it reconfigures them.

In 2019, Milano Arch Week sought to reflect on the challenges posed to architectural and urban language by the ongoing environmental transformations. The same happened with Design Week and Fashion Week. Virtually all the city's major "creative" events have become opportunities to experiment with languages and forms that allude to the ecological and climate crisis. These events have generated responses anchored in specific knowledge, aimed at different audiences.

Communicating the crisis requires more than publicising figures: it requires putting scientific data and speculation about its consequences into sensible forms, i.e. inscribed in the present. This means that there will be multiple ways of translating the climate event into forms, objects and technologies that make up the urban environment and generate new social practices.

The intelligibility of the urban forest, therefore, should not be sought in an abstract ideal of sustainability, but in the way it is articulated to the concrete conditions of the Milanese environment—its materials, its institutions, its publics and its practices.

But why start with these brief stories about design and architecture? These responses have been formulated within architectural and urban planning practices that help to explain the characteristics that this transformation has taken on. As part of growing professional specialisation, contemporary urbanism has been guided by approaches that prioritise the quantification and visualisation of climate phenomena in terms of a territorial scale. While this quantification is undoubtedly important, it is rarely accompanied by a deeper questioning of what social practices are generated and what forms of agency are activated in the territories. Such questions seem to escape the predominant analytical devices in the field of urbanism.

In an interview with the Italian newspaper Corriere della Sera, Stefano Boeri reinforced this argument, pointing directly to the need to transform the ways in which urban space is produced:

> "Only from the urban fabric can we find concrete solutions to tackle global warming. That's why we need to bring nature back to cities. The Vertical Forest—with all its ramifications—is nothing more than an attempt to make metropolises greener. Where it's not possible to intervene with large parks—like Central Park, with its 20,000 trees, created in 1857—we can design vertical, super-concentrated forests".

After each episode of "bad weather" that hit the city, the newspapers reported the damage with an emphasis on the immediate inconvenience—which irritated activists who preferred to see these events contextualised within the broader framework of earth system transformations. As M. noted in an exchange of messages: "Words are important. Continuing to call it 'bad weather' legitimises inaction, because it gives the impression that it's something cyclical, inevitable - 'it was yesterday, it is today, it will always be'. Talking about the effects of climate change imposes responsibility" (23/10/2019—chat).

A report by Legambiente reinforces this perception: between 2010 and 2021, Milan recorded 30 significant episodes of extreme weather events, making it one of the Italian urban areas most impacted by these occurrences.[4] But the tone of the narratives is one of appreciation for the direction Milan is taking. It is necessary to protect itself "without losing attractiveness, without excluding growth and without harming business" (Reinventing Cities event—Triennale/2019). In this sense, the city administration's participation in the 100 resilient cities and C40 networks is seen as a positive sign:

> "No subject is closer to my heart", swears the mayor. He recalls his commitment to the "C40 Cities" network, the association that brings together the world's largest cities. There, he explains, "we are all extremely

[4] https://www.legambiente.it/wp-content/uploads/2021/11/Report-OsservatorioCittaClima2021.pdf.

> convinced that this is the most sensitive issue for the future of our communities. Unfortunately, it's also the most neglected. Everyone says they are aware of the problem, but few are willing to make their own sacrifices. (…) For Lombardy's centre-right, the solution is already in place. "Paolo Grimoldi, regional secretary of Salvini's party, replied: "Let's speed up the road to autonomy. "Let's do it quickly, so that an organised region like ours can have more means to monitor its territorial areas with the greatest hydrogeological risk and prevent and limit the damage caused by bad weather, which, with global climate change, is no longer an isolated exception".[5]

One of the results of these long-term strategies was the emergence of the Forestami project. Created in 2018 as the result of a working group at Politecnico di Milano, the project aimed to study ways of intervening in the tree component of the city of Milan and its metropolitan area. Forestami is promoted by an institutional coalition made up of the Metropolitan City of Milan, the Municipality of Milan, Parco Nord Milano, Parco Agricolo Sud Milano, ERSAF (Ente Regionale per i Servizi all'Agricoltura e alle Foreste) and the Milan Community Foundation.

The plan calls for 3 million trees to be planted by 2030, with the aim of improving air quality, combating the effects of climate change and improving urban life in Greater Milan. Initially conceived as applied research by the Politecnico, the project was realised with the financial support of the Falck Foundation and FS Sistemi Urbani (a state-owned railway infrastructure company), making it a public–private partnership.

In addition, Forestami is promoting crowdfunding campaigns aimed at the Milanese population, allowing citizens to make direct donations—even in the symbolic form of a gift. The campaign is accompanied by the slogan: "The best gift is to contribute to the growth of your city's natural capital. Birthdays, graduations, or simply to celebrate a special moment".[6]

The aim of the project was to identify "measures to increase urban vegetation (in its various forms: from afforestation to green roofs and walls) and drainage surfaces to respond to rising summer temperatures and reduce the risk of flooding from extreme weather events" (ACP: 118). The municipality, however, is not alone. Several others have joined

[5] https://milano.corriere.it/notizie/cronaca/18_novembre_05/allerta-maltempo-sfogo-sala-rischio-climatico-adesso-basta-98128f6e-e0eb-11e8-b7b1-47f8050d055b.shtml.

[6] https://forestami.org/regala/.

the large Forestami programme, as has the Lombardy region itself, which since 2017 has been funding the construction of green infrastructure, the aim of which is to "protect and improve the environmental system of the rural landscape through the creation of new forests and complex green systems".[7]

From landscape planning to ecological restoration of ecosystems, there are many ways in which the forest enters the calculations of urban planning. As Myers notes, we need to "understand the ways in which garden designs inform, sediment and disrupt the very infrastructure of relationships between plants and people, especially in cities" (2019: 116). The urban forest must not only enrich the forms of ecological interaction in the territory—promoting resilience, protection and biodiversity—but also respond to the challenges of atmospheric control and CO_2 absorption generated in the urban space itself. In the urban fabric, the tree is reconfigured as a clean technology for human and non-human well-being.

The urban forest can take many forms, and it would be a mistake to reduce it to a single function (recreational, aesthetic, environmental). It is always part of a whole—made up of intentional interactions (the object of urban design) and unintentional ones (spontaneous rhizomatic expansion). In this sense, the forest is a device for redesigning both urban and climatic aspects and everyday practices on a micro-scale, in the interactions between high- and low-carbon materialities.

When transformed into multifunctional biotechnology, the urban forest requires a socio-spatial dimension that makes it possible to "socialise" plant beings: to create spaces in which they can associate, proliferate and transform relationships between humans and non-humans. Trees and vegetation need to be inserted into new semantic and material fields (Battaglia 2017), articulated with socio-technical imaginaries about the future of society and the earth system. Their concrete form will depend on their relationship with existing socio-materiality—they can either adapt the business-as-usual of high-carbon socialities or open up space for new infrastructural possibilities (Nolan 2018; Carson 2019).

This is the case of the urban forest as a " technologically enhanced garden-in-the-making, built by workers interpreting architectural designs. This was a garden that would have to be tended, weed, and watered to keep it thriving" (Myers 2019: 116).

[7] https://www.bandi.regione.lombardia.it/procedimenti/new/bandi/bandiByTitolo?titolo=Infrastrutture+verdi.

Trees have historically been codified according to their capabilities. As living beings, they interfere with the living conditions of other beings, both biotic and abiotic (Coccia 2018). This presents an analytical challenge for the social sciences, especially in urban studies, where materiality cannot be ignored, but is often treated as an effect of political and economic structures. In this way, the forest tends to be instrumentalised: an accessory to a political project or vision of the future (Angelo 2019a).

However, we should recognise that the design of urban forests shapes possibilities and blocks others, acquiring political performativity. Its functionalities range from improving quality of life to producing what Nixon (2011) calls "slow violence": from eviction through price rises to green gentrification.

Tsing (2015) identifies the plantation as the model that allowed industrial processes to scale up. James Scott (1998), for his part, identified the emergence of forestry as a paradigm for state rationalisation of the territory. The forest had to be transformed from a "disorganised" whole into a stable and manageable system, reduced to the production of wood and other essential resources. This rationalisation involved simplifications: selecting species, classifying attributes, excluding what was considered pestilential or unviable.

The "social engineering" promoted by forestry aimed to create a manageable environment, a landscape moulded by principles of predictability. The planted forest entered a circuit of different codifications: resource, energy, aesthetics and heritage. To this end, uniformity was key: the more homogeneous, the more controllable. Theory merged with practice, creating a form, a trajectory and a classificatory structure.

But urban and forestry planning practices are changing, precisely through the intrusion of agents (human and non-human) that destabilise rational planning. This doesn't mean abandoning scalability, but redefining it. Resilience emerges as a way of integrating stress and disturbance as normal aspects of the system.

The role of urban planners and architects is justified by the task of massively installing these plants in heterogeneous environments, in which trees must coexist with buildings, cars and infrastructures, each generating specific affordances (Easterling 2021) for plants, animals and humans.

The urban forest becomes a device—made mostly in urban scale by urbanists and architects—whose goal is the *infrastructuraction of a protective milieu to resist extreme weather events*. The relationship between the climate crisis and green infrastructure can be read as an injunction to create new chains of dependency under the banner of security. But it can also be seen as an opportunity—as we'll see in Chapter 8—for the formation of multi-species collectives that escape the fossilised socialities gestated in and by the "Great Acceleration".

Thus, situated as a biotechnology that rehabilitates anthropised environments, the urban forest not only materialises, but also generates different political perspectives on how to respond to Gaia's insistence on our unstable and vulnerable collectives. Leigh Star's distinction between background and foreground remains pertinent, as it allows us to observe how infrastructures become visible or invisible depending on the relationships in which they are inserted, the uses they receive and the practices that mobilise them. In this sense, the urban forest is simultaneously a technical system, a living ecosystem and a political artefact.

4.2.1 New Agents and the Redesign of Urban Assemblages

The climate becomes the object of explicit techniques linked to the socio-material and energetic conditions of urban space. Within this logic of environmental governance, the forest—as a plant, vegetation or green cover—takes centre stage. Urban vegetation returns as a thermostatic regulator and protective barrier, articulating itself as ecological infrastructure.

In the Aria Climate Plan, trees are positioned as key elements of the climate adaptation and mitigation strategy. According to the document, they must:

"Mitigate heatwaves and floods, provide the city with new ecosystem resilience services to reduce the 'heat island' effect; reduce air pollution (30 $\mu g/m^3$ of PM 2.5 - three times the WHO safety limit); reduce energy consumption linked to the use of air conditioning, contributing to the target of an 80 per cent reduction in greenhouse gas emissions by 2050 (net zero emissions, C40 cities); ensure social inclusion and cohesion through community projects aimed at regenerating peripheral areas; expand green and blue infrastructure and their ecological connections; significantly increase the number and biodiversity of plant and animal species" (ACP 2021: 801).

The technical language of mitigation and adaptation thus finds in trees and urban vegetation a means of materialising immediate responses to the climate crisis, converting the urban forest into a real infrastructural device—with a dignity (albeit contested) comparable to that of road or hydraulic infrastructures.

Milan's climate strategies should therefore be understood as localised experiments within broader models of climate urbanism (Goh 2015; Blok 2014; Blok & Meilvang 2015), spread by international networks in which the city is an active participant.

However, it is essential to recognise that the presence of urban forests does not play a univocal normative role. As Marres (2015) reminds us, the agentivity of plant beings will depend on the associations to which they belong. Their political, ecological or symbolic effectiveness is relational and situated. Pellizzaro exemplifies this logic when he states:

> "Forests are a solution, they are lungs that help absorb pollutants. But in an urban context, it can be much more effective to plant 10 trees in Via Vittor Pisani (the avenue leading to the central station, surrounded by buildings) than 10,000 in Parco Nord. These 10 trees on Via Vittor Pisani reduce the ambient temperature for pedestrians and pavements, cast shade on the surface and on the surrounding buildings, improving their energy performance".

The question, therefore, is not just how people (residents, workers, tourists) perceive urban greenery, but how they interact with it. Not just what they do with the trees, but also what the trees make them do.

As I have tried to demonstrate throughout this research, this articulation can range from the symbolic gesture of the mayor of Milan planting a tree at an institutional event to the political action of activists in occupied social centres doing the same. The forest, in this context, becomes a point of intersection between multiple forms of agency, infrastructure, political imagination and everyday practice.[8] If in the first case trees can be inserted as functional components in the local ecology and climate, in the second they obviously can't be restricted to this function: they could be defined in the dynamics of an "antagonistic political forestry" that tries to break the cementation of the territory—done without authorisation and in open defiance of local urban politics; something in which FFF-MI activists decide, in a demonstration, to plant a tree in Baiamonti square without authorisation from the municipality.

[8] https://www.facebook.com/mutuosoccorso/photos/4120398171420256.

The urban forest is defined by the municipal administration, together with other institutions, as part of a green infrastructure. And the way this has been codified today—see, for example, the European Commission's working group.[9] Infrastructure can be defined as a network of distributed areas in a territory capable of providing a wide range of services. For the EU Commission,

> "Green infrastructure is a strategically planned network of natural and semi-natural areas with other environmental features designed and managed to deliver a wide range of ecosystem services such as water purification, air quality, space for recreation and climate mitigation and adaptation."

Despite the functional coding that structures urban afforestation policy, it is possible to say that each point in this network—each initiative or plant intervention—will be articulated in a particular way with the territory in which it is located. A peri-urban green area will not play the same role as a green area located in the city centre. The performativity that each of these interventions produces on human and non-human agents will be different.

This pluralism doesn't just refer to the diversity of material forms or "ecological goods" generated, but emerges from the empirical multiplicity of links, uses and interpretations. There are geographical specificities (centre versus periphery), different uses of existing devices (such as parks) and new urban platforms (roofs, façades, balconies, streets). In addition, trees will be involved in multiple problematic situations: they will serve as part of political strategies, as examples of architectural innovation or as a symbol of corporate responsibility.

Defining the urban forest as infrastructure offers a useful analytical key to follow how urban relational space is constantly redefined—and how non-human agency is territorialised in this process.

In the specific case of the urban forest, the boundaries between "infrastructure" and "environment" become, as Carse (2019: 102) states, " artefact of the constant organisational work (discursive, scientific, technical, physical) required to make and maintain connections in a living, encroaching world". This is also where the institutionalised forest differs from the practices and visions cultivated by activists.

[9] https://ec.europa.eu/environment/nature/ecosystems/index_en.htm.

"Urban forestry" is a generic term that translates in practice into an increase in vegetation cover at multiple scales: from public parks to building interiors, passing through areas formally defined as urban or rural. The process of afforestation seeks to constitute an infrastructure—in that it presupposes an articulated, interdependent and planned operation, whether public or private.

As the Piano di Governo del Territorio (PGT 2019: 35) emphasises:

> "Understanding how global climate change influences the local urban climate, highlighting the areas most exposed to the phenomenon of heat islands, or monitoring the evolution of soil consumption, identifying the most impermeable areas of the city where the positive effects of green roofs would be most noticeable".

As Carse (2019: 103) notes, " nature can become infrastructure through concerted investment and management—that is, infrastructural work". The Aria Climate Plan working group was guided by this logic, proposing "significant and targeted vegetation in more urbanised areas, to cool the zones most vulnerable to heat waves, as well as peri-urban forestry measures in underdeveloped zones (agricultural and/or residual), with the aim of strengthening a green belt capable of reducing the temperature of the urban system as a whole" (ACP: 120).

This is a striking example of what Marvin defines as an atmospheric disconnection from the meteorological commons; for Stephen Graham (2018: 251), " one future focus of urban design on a urbanising and warming planet should be the weather itself".

The Forestami project exemplifies how urban planning translates the simultaneous challenges of the climate and ecological crisis into specific material forms—while at the same time redesigning local socio-material ensembles, including their meteorological, hydrological and atmospheric conditions. Considered ambitious even by international authorities—as stated by the representative of Bloomberg Philanthropies in New York at a Triennale event—the project aroused different reactions: bewilderment on the part of those who saw car parks disappear, enthusiasm among those who saw it as a possible solution to the problems of urban pollution.

Forestami proposes creating a new urban ecology, based on the massive introduction of trees and their association with everyday practices, both private and collective. Among the stated objectives are: mitigation of the heat island effect, reduction of fine particles in the air, thermal regulation, hydrological security (ACP: 760).

By planting trees, the aim is to influence energy consumption habits (reducing the use of air conditioning), leisure practices, urban mobility, voluntary engagement and even aesthetic experiences of everyday life. The urban forest, in this sense, is understood as a means through which new forms of sociality could emerge. As the programme states:

> "Multiplying the number of plants along streets, squares and courtyards, on the roofs and facades of our homes - is the most effective, economical and engaging way to slow down global warming, reduce energy consumption, clean the air we breathe and improve the well-being of citizens".[10]

The forest, as a technique for composing protected spaces designed to guarantee the enjoyment of optimised ecosystem services, has become an obligatory device in territorial regeneration and urban development projects in Milan—particularly in emblematic initiatives for the city's future, such as the projects selected in the Reinventing Cities programme (C40 Network), which include areas such as former railway stations totalling more than 1 million m^2.

The urban forest thus becomes legible from the new urban complex under construction. As experimental as it may be, this new infrastructure must be situated within the movement that, according to the criticism of Milanese activists, combines "property speculation and greenwashing" (D., 30, engineer). However, this critical judgement deserves to be deepened. As argued above, the forest is consolidating itself as an urban planning device, offering a new meaning to the proposal of ecological urbanism. It doesn't act as a limit to urban development—on the contrary, by integrating itself into it, it reconfigures its contours.

The public issue that emerges from the urban forest offers an opportunity to observe the design process as a "work in progress", permeated by controversies over how to deploy this infrastructure in a complex and unequal city. Christian Nold (2018: 97) proposes a relevant analytical key when he talks about an "active methods of infrastructuring" where the target is "political assembling around an issues".

[10] https://forestami.org/obbiettivo-2030/.

In this context, biodiversity is brought into the centre of cities, not just as a green space for leisure, but as a condition for the installation and permanence of other species. Even so, the presence of these plant agents begins to disrupt traditional planning schemes. The location of trees becomes a matter of dispute, mobilising divergent voices—both supportive and critical. There are growing warnings about the possible gentrification of these areas, with rising rents and property prices, making it impossible for low-income populations to stay.

Despite the scale of the project, critical analyses of its implementation rarely find space in the mainstream media. This role has largely been taken on by activists. These collectives have endeavoured to expose Forestami's internal contradictions, revealing oscillations between declared objectives and the means actually used.

In an article published on the FFF-MI (Fridays for Future—Milan) website, Andrea Ascari intertwines the history of Forestami with the trajectory of local committees fighting to preserve green areas and combat the advance of soil sealing. According to him, a "common feeling" is being consolidated: the appreciation of urban trees combined with the demand for large green areas that provide cooling, CO_2 absorption, drainage and rainwater reabsorption—and, why not, also aesthetic value.

However, Ascari warns that, in the end, "the trees often become bushes and thickets, with no guarantee of maintenance or irrigation, which means certain death in Milan's scorching summers". And meanwhile, he says, "the city council continues to cut down hectares of trees, like in Bresso and Goccia, or threatens to 'reform' parks like Ticinello". For him, Forestami's real challenge should have been to "remove asphalt and concrete to create large new green areas, without building more".[11]

Shopkeepers are also raising their voices to protest against the possibility that the introduction of trees in shopping areas could harm the flow of motorised consumers. The proposal to plant trees in Corso Buenos Aires, a shopping area in the city centre, was rejected by 72% of the 127 commercial operators in the area. According to the president of Ascobaires and the Rete associativa vie Confcommercio Milano:

> "The vast majority of Corso Buenos Aires operators have clearly expressed all our concerns: the elimination of car parking, which is fundamental to Corso Buenos Aires; the widening of pavements that were widened a few

[11] https://fridaysforfuturemilano.it/i-comitati-fanno-rete/.

> years ago; the certain risk of neglect: just look at the sad spectacle of wilted plants in Via Vittor Pisani. I would also add the total lack of maintenance on the streets near the Corso, which have been abandoned to decay. On the contrary, we are thinking of proposals to further improve Corso Buenos Aires and its commercial attractiveness: wi-fi, street furniture with street lamps, possible flower beds, but well looked after".[12]

Much has been discussed about aspects that reveal the precarious position of the multiplicity of plant agents involved in the urban forest, especially in relation to the issue of maintenance. This aspect, however, seems largely absent from the institutional formulation of the project. When it is mentioned, it is resolved by hiring specialised labour or by the expectation that "social" forms of maintenance will be generated in the very process of setting up the urban forest—in other words, responsibility for care is partly delegated to the community dynamics that form around planting.

4.3 The Green House as a Living Machine

The proposal to plant three million trees in the metropolitan area of Milan not only gave rise to a variety of imaginaries about the urban future, but also fuelled the proliferation of experiments based on this imaginary. The afforestation proposals managed to mobilise part of the population, but the way in which these citizens appropriated the idea generated practices that, in some cases, were in tension with the official model promoted by the city administration.

This is not to say that these alternative practices didn't exist previously. On the contrary: the point is to investigate the frictions that emerge from the formalisation of the programme and its implementation, especially when confronted with ways of planting and caring that don't align with institutional logic. The urban forest has thus become a privileged topic of public debate, which has given rise to a range of activities—from scientific meetings and trade fairs to educational workshops and artistic exhibitions. Plant beings have acquired a new political and symbolic status in this process.

Of course, this recognition of plants as life-sustaining agents did not come about suddenly. It is part of a wider movement to reimagine and

[12] https://www.greenplanner.it/2019/04/08/progetto-corso-buenos-aires-milano/.

reconfigure urban spaces, which has accompanied both planning practices and urban activism in recent decades (Scott 2016).

Our focus here is on what I defined in the first chapter as turning explicit an social environment—a process that does not require the formulation of a single, objective perspective, but rather the articulation of multiple visions and practices on how to tackle the problems brought about by climate change, including the issue of non-human agency.

The Municipality of Milan has funded dozens of projects aimed at the climate agenda. Short-term workshops and long-term initiatives were organised on the basis of priorities defined by the analytical framework drawn up by the administration. Between 2018 and 2021, I followed several of these initiatives, in which trees were presented as an adaptive solution to the climate crisis. They were mobilised as: nature-based solutions; home adaptation devices; instruments to contain urban sprawl or break up the concrete grid; regenerative alternatives for peripheral or degraded areas.

One of the most visible consequences of this green turn has been the reconfiguration of domestic spaces as environments conducive to plant life. According to the specialist portal Agricommercio e Garden Centre, "in 2019, the value of Italian flower production exceeded 2.7 billion euros. After a decade of retraction in the sector, the data showed growth for the third consecutive year: +160 million in value in 2018, +176 million in 2017".

This data points to the way in which the symbolic flourishing of urban vegetation is linked to production chains, commercial circuits and forms of consumption that are not dissociated from the material and political transformations taking place.[13] Another website specialising in interior design pointed out the same trend:

> "Houseplants are selling like hotcakes in large metropolises, reflecting a dominant trend in favour of biophilia, i.e. reconnecting with nature. What's more, the micro-gardening craze is testimony to the lasting changes in society, as well as the growing consumer focus on well-being, home and autonomy".[14]

[13] https://agricommerciogardencenter.edagricole.it/giardinaggio/il-florovivaismo-italiano-fino-al-2019-cresceva/.

[14] https://nellyrodi.com/en/micro-gardening-the-new-lifestyle-craze/.

And so Corriere della Sera itself, in its Living section, reported on this new trend in the domestic sphere:

> "At a time when people are spending much more time at home, alone or with other people, the micro-gardening trend has literally exploded. To be fair, even before the pandemic, the phenomenon was undergoing a major revival, especially in big cities where there is less greenery".[15]

In times of pandemic and climate crisis, the proliferation of domestic solutions seems to be a specific response to the anxiety caused by the event. I don't exactly see a discontinuity between the resilient city and domestic spaces adapted to withstand climatic conditions that are not found outdoors. In a study that attempts to synthesise different research on the performance of plants in domestic spaces, Virginia Lohr (2010) pointed out four important aspects, three of which could be considered with a high degree of certainty: firstly, the reduction of air pollution; air humidity; the reduction of particulate matter and acoustics, where the data did not allow us to conclude on their effectiveness.[16]

The way in which the domestic space is articulated with the chains of dependence is just as relevant as the dynamics of public spaces. Daily consumption practices are reconfigured based on people's status as climate citizens. This implies continuous attention to daily habits that can perpetuate patterns of high consumption: switching off lights, avoiding food waste, replacing plastic bottles, reducing meat consumption.

These seemingly banal gestures are reinterpreted as expressions of new skills acquired at a personal level to live in an era marked by the permanent climate emergency. It's about incorporating solutions made possible by devices and technologies available on the market, which are part of a grammar of individual responsibility and active participation in the ecological transition.

At this point, a central aspect of the resilient city and its socio-technical devices emerges: they presuppose the existence of certain skills—or require the acquisition of others. In order to move around without cars, for example, or to change their eating habits, individuals need to be able

[15] https://living.corriere.it/tendenze/green/piante-online-micro-giardinaggio-millennial/.

[16] https://public.wsu.edu/~lohr/pub/2010LohrBenefitsPltsIndoors.pdf.

to meet the new requirements that define the behaviour expected of a climate citizen.

In the case of urban mobility—shared bicycles or scooters—you need to have a smartphone and know how to operate it to access the apps. In the case of food, this involves additional costs, access to alternative distribution circuits and the availability of time to avoid processed or long-lasting food. These are even the three main difficulties identified by citizens themselves in a survey carried out by the municipality.[17]

The investment in the domestic space as an adaptive module offers important clues as to how to build a resilient city. Although I cannot explore here all the aspects implemented in the domestic sphere—such as the use of renewable energies, smart meters, purification filters, automation devices, among others—it is possible to state that the home environment has become the object of an explicit climate design, whose relationship with the environment is particularly revealing (Sloterdijk 2015: 519).

In addition to commercialised architectural solutions, the housing module can be structured in such a way as to effectively reduce energy consumption and CO_2 emissions. However, this requires a number of technical and technological innovations. As Marres (2015: 127) points out, the ecological house:

> "can also be understood as a device for performing the comprehensive kind of change that is involved in the re-ordering of social, material, technical, economic, political-and-son-on relations. Eco-homes are configured as devices for technological changes, the transformation of lifestyle, the democratization of environmental governance, the exploration of alternatives culture and so on."

The house designed to be low-carbon is part of the responses to climate change implemented individually, as a family or by small residential collectives. The risk of fragmentation of the unitary city (Graham & Marvin 2001) and the emergence of resilient enclaves (Hodson & Marvin 2010) take on a more granular dimension here: it's not just a question of territorial differentiation between neighbourhoods, but of

[17] Some data can be accessed in the survey carried out by the municipal administration. https://www.comune.milano.it/documents/20126/126287263/05102020_Cittadini+e+cambiamenti+climatici_un%27indagine+esplorativa.pdf/c89cdaa0-41dd-fadb-b0d0-06cecfd7da9b?t=1602067034807.

inequalities that manifest themselves in the very basic units of the urban landscape—the house.

Opting for a low-carbon lifestyle in the domestic sphere comes at a cost. In a scenario where the price of energy and fossil fuels cascades to the price of other goods, the domestic adaptive module becomes a concrete expression of a struggle for habitability—understood here as the material guarantee of subsistence in a given territory. The home, as a private environment, can open up to the outside world (by connecting to public or community initiatives) or close itself off in circuits of consumption and dependency. As we have seen, it is not a question of looking for a general theory of domesticity, but of understanding how the domestic space articulates with this "outside".

According to Knox (2020: 199), what the residents "were doing with ecohouses, even in the most experimental setting, was not creating universal truths but rather deconstructing, or disrupting, objects in order to reconstruct them anew - inhabiting the house thus becomes an opportunity to relearn what a house is in thermodynamics terms (…) allow houses to be transformed into techniques of political intervention".

This reading helps us understand, for example, the uncomfortable reaction of an association representative present at a home gardening workshop when she heard reports about planting practices carried out by activists in occupied public areas. The emerging tension was quickly ended by her with the phrase: "everyone does what they want". This moment reveals a fundamental question: What kind of public emerges from the experience of becoming active within domesticity? What forms of climate policy are generated from the creation of these adaptive modules?

Although I can't answer these questions exhaustively, the case observed shows that the response to climate change has been formulated here from a perspective that avoids direct confrontation with institutions or companies. As Simona said: "We can't wait for political decisions to take action. The state has already shown itself incapable of responding with the necessary urgency". Although criticism of the role of large corporations—responsible for "poisoning the planet"—was voiced, the way out was to build a certain autonomy from these chains of dependence. In this context, the workshop became an opportunity to explore how the idea of autonomy is articulated in practice by subjects who want to "become active".

The meeting was not just a time for technical learning about growing vegetables at home, but a practical lesson on how to create an adaptive module within the domestic unit, aimed at self-sufficiency in a "toxic environment" (Simona).

One of the most significant consequences of the spread of green solutions is their transposition into spaces not originally planned. As a result, many plants have left planned public spaces and moved into domestic environments. Several informants reported changes in the configuration of their homes, adapted not only to accommodate plants, but to turn them into agents that transform habits and microclimates.

During the workshop, Paolo (42) told me that he had recently started a small garden on his balcony. According to him, "it was a way of exercising my right to eat what I grew myself and to know what's in it". When I asked him if this would guarantee food sovereignty in the face of industry, he admitted that it wouldn't, but emphasised that the practice made him "less dependent" and gave him pleasure: "I like knowing that I'm eating something I've grown myself".

The same happened with her friends Anna (35) and Stefania (37), who lived together. They had taken gardening courses and were growing different species inside the flat and on the balcony. "We got closer to nature", they said. Anna commented that the plants reminded her of her trips to Indonesia and Brazil, "where nature was abundant". Alice said that, thanks to the plants, she began to realise how grey "the city outside was", and how "it needed more trees, greener, more life".

Plants are also distributed at public events and, in some cases, offered for temporary adoption before permanent planting—as promoted by several community gardens in Milan. These small gestures reveal how the practice of home gardening can function as a device of ecological subjectivation, reconfiguring not only the interior of homes, but also the way their inhabitants position themselves in relation to the city and the climate crisis. The workshop to build a "bioactive vegetable garden" was one of the activities of the project called "Forchette verdi e milano a tavola col Pianeta".[18]

The event was held on 20 October 2020. Simona, the creator of the project—funded by the Municipality of Milan—made her intentions clear: "every individual can contribute to reducing environmental impacts and the effects of climate change". The project consisted of creating an app (Wiseair) to measure the ecological impact and emissions linked

[18] https://www.produzionidalbasso.com/project/forchette-verdi-E-milano-A-tavola-col-pianeta/.

to consumption practices. This would be a way of "doing something without waiting for governments". The second initiative was to transfer these sustainable food choices to the creation of community gardens on balconies and terraces, using bioactive garden techniques, which guarantee high-quality nutraceutical vegetables, without chemicals, with low water consumption, self-healing fertility and greater CO_2 sequestration in the soil. The third initiative was to use Wiseair technology to monitor air quality and thus "close the circle".

According to Simona, "food was the most immediate way of doing this on our own". The project aimed to influence food choices and transform the home itself into a potential food producer and environmental monitoring point. It is a paradigmatic case to discuss how many people have begun to introduce plants with the aim of modifying the composition of the air in their internal environments, establishing symbolic and material relationships with these plant agents and seeking partial food self-sufficiency, disconnected from industrial chains.

This introduction of plants is part of a speculative design, oriented towards the interaction between material, technological and biovegetal agencies, which Battaglia (2017) calls a biotechnological assemblage—with the aim of emphasising well-being and domestic autonomy. The project is part of the public calls promoted by the city council, aimed at small, short-term interventions that encourage environmental awareness practices. However, the way in which these activities are appropriated by the participants can vary considerably, escaping institutional foresight.

One of the motivators for the project was "frustration at how little progress has been made towards the sustainable millennium goals". For Simona, it was essential to "inform and inform better", because "changing the way we eat and agricultural practices could reduce greenhouse gas emissions by up to 50 per cent". The garden was conceived as a device for producing healthy food and capturing CO_2, with the potential to expand beyond the private domain.

During the event, there was a practical demonstration on how to set up a bioactive garden. Luca, organiser and member of the Confederazione Cittadinanza Consapevole, showed how to prepare a container with clay (2 cm), soil (40 cm) and plant seasonal species, such as cicoria rosa and cima di rapa, donated by members of a solidarity consumer group (Gruppo di Acquisto Solidale).

According to him, this practice has multiple impacts: job creation, the production of healthy food and the dissemination of critical knowledge about the greenwashing promoted by large distributors through organic labelling. This is an experience with potential for expansion and reproduction.

The group present was small and heterogeneous: 12 people, all Italian, mostly women and over 35. The event ended with a vegetarian cocktail. Despite the pandemic context, the common desire among the participants was to find the technical means to make a path of affective-material transformation possible. Growing their own food was a way of asserting autonomy from the industrial agri-food chain.

The domestic space has come to be understood not just as a receiver of resources and emitter of externalities, but as a place of active experimentation. If the house, as Le Corbusier defined it, is a "machine for living", then its material and symbolic modifications offer a concrete input for imagining housing futures under new ecological conditions.

The houseplant thus joins the trees of urban afforestation as part of a wider green infrastructure. Another workshop observed took place in an apartment block in Via Padova, an area of Milan characterised by its migrant population. The proposal was to involve the building's residents in a community garden, but only children attended. The adults present came from other parts of the city.

The meeting was led by a member of the *Orti di* Via *Padova*, who explained the technical procedures: choosing a pot, preparing the soil, selecting species and maintenance. The reasons for cultivating varied: beauty, company, food, socialising. Vegetation, in this context, becomes malleable to pragmatic re-significations, and can be incorporated into existing arrangements or become a vector for new ones.

What happens in the domestic domain depends on what happens outside it. Instead of generically evoking the "context", we should consider the environment that is being actively constructed. The main distinction between the perspectives is perhaps here: for planners and managers, space is delimited by internal functionalities and walls that guarantee isolation; for activists, it is an expanded space that includes the ecological footprint in distant territories—rejecting, for example, the exploitation of remote ecosystems to produce consumer goods.

This difference can be seen in the varied (and not necessarily opposing) forms of urban redesign and the relational matrices between humans and non-humans. Our aim was to show how this world is being composed:

how people live, how public and private spaces are constructed within what I call low-carbon socialities, where non-human agentivity is not excluded, but inserted into specific relational spaces.

Autonomy, after all, also has its risks. The availability of technical solutions to enable the performativity of a low-carbon lifestyle speculatively configures one of the scenarios for the near future—one in which autonomy is sought through dissociation (partial, always) from the degrading ecological and climatic conditions of the immediate environment.

Climate disaster bunkers offer, like the space stations described by Sloterdijk (2015: 299), an extreme image of a world that has made explicit the work of designing survival conditions in the face of atmospheric collapse and local degradation (Graham 2018: 340).

The forest is just one of several devices mobilised to transform the urban machine of the Anthropocene into a resilient urban system—or at least to propose this transition. By being described as an investment capable of protecting and attracting new flows, urban space is reterritorialised by the systematic measurement of all its material and immaterial components. However, the transformation of the city into a forest also frees up lines of flight (Deleuze & Guattari 1987), through which subjects glimpse possibilities of breaking away from extractive meshes. To become other, on a smaller scale, is to become active with the plants, against the cementation of the urban machine.

References

Angelo, Hillary. (2019a) Added value? Denaturalizing the "good" of urban greening. *Geography Compass*;13:e12459

Angelo, Hillary. (2019) The greening imaginary: urbanized nature in Germany's Ruhr region. *Theory and Society*; 48: 645-669.

Battaglia, Debbora. (2017) "Aeroponic Gardens and Their Magic: Plants/Persons/Ethics in Suspension." *History and Anthropology* 28 (3): 263–292. https://doi.org/10.1080/02757206.2017.1289930

Blok, Anders. (2014) Worlding cities through their climate projects?, *City*, 18(3), pp. 269-286. https://doi.org/10.1080/13604813.2014.908514

Blok, Anders; Meilvang, Marie Leth. (2015) Picturing urban green attachments: civic activists moving between familiar and public engagement in the city. *Sociology*, vol.49(1): 19-37.

Carson, Ashley. (2019) Dirty landscapes: how weediness indexes state disinvestment and global disconnection. In: Hetherington, Kregg (ed.). *Infrastructures, environment, and life in the anthropocene*. Durham: Duke Press University.

Coccia, Emanuele. (2018) *The life of plants*: A Metaphysics of Mixture. Cambridge: Polity Press.

Comune di Milano. (2019) Piano di Governo del Territorio (PGT) 2030: Documento di Piano, Piano dei Servizi e Piano delle Regole. Approved by City Council on October 14, 2019; effective from February 5, 2020. Milan: Comune di Milano.

Deleuze, Gilles and Guattari, Felix. (1987) *A thousand plateaus*: capitalism and schizophrenia. Minneapolis: University of Minnesota Press.

Easterling, Keller. (2021) *Medium design*: knowing how to work on the world. London: Verso.

Gabrys, Jennifer. (2020) "Sensing a Moving Planet." In *Critical Zones*: the science and politics of landing on Earth, edited by Bruno Latour and Peter Weibel, 254–263. Cambridge, MA, and Karlsruhe: MIT Press and Zentrum für Kunst und Medien.

Goh, Kian. (2015) *A Political Ecology of Design*: Contested Visions of Urban Climate Change Adaptation. (Thesis) Urban and Environmental Planning, Yale University.

Graham, S. and Marvin, S. (2001) *Splintering urbanism*: networked infrastructures, technological mobilities and the urban condition. London: Routledge.

Graham, Stephen. (2018) *Vertical*: the cities from satellites to bunkers. London: Verso.

Hodson, Mike, and Simon Marvin. 2010. "Urbanism in the Anthropocene: Ecological

Knox, Hannah. (2020) *Thinking like a climate*: governing a city in times of environmental change. Durham and London: Duke University Press.

Lohr, Virginia I. (2010) "Plants and the Individual: A Recent History." *Journal of Home & Consumer Horticulture* 1 (2–3): 225–230.

Marres, Noortje. (2015) *Material participation*: technology, the environment and everyday publics. London: Palgrave Macmillan.

Mostafavi, Mohsen; Doherty, Gareth (eds.). (2016). *Ecological urbanism*. Cambridge: Harvard University Graduate School of Design.

Myers, Natasha. (2019) From Edenic apocalypse to gardens against Eden: plants and people in and after the Anthropocene. Hetherington, Kregg (ed.). *Infraestructures, environment, and life in the anthropocene*. Durham: Duke Press University.

Nixon, Rob. (2011) Slow Violence and the Environmentalism of the Poor. Cambridge, MA: Harvard University Press.

Nold, Christian (2018) Turning controversies into questions of design. In Marres, Noortje; Guggenheim, Michael; Wilkie, Alex. Inventing the social. Manchester: Mattering Press.

Scott, Felicity (2016). *Outlaw territories*: environments of insecurity/ architectures of counterinsurgency. New York: Zone Books.

Scott, James. (1998) *Seeing like a state*: how certain schemes to improve the human condition have failed. New Haven: Yale University Press.

Sloterdijk, Peter (2015). *Sfere III*: schiume. Milano: Raffaello Cortina Editore.

Stépanoff, Charles. 2024. Attachements. Enquête sur nos liens au-delà de l'humain. Paris : La Découverte.

Stoetzer, Bettina. (2018) "Ruderal Ecologies: Rethinking Nature, Migration, and the Urban Landscape in Berlin." *Cultural Anthropology* 33 (2): 295–323. https://doi.org/10.14506/ca33.2.09

Tsing, Anna. (2015) *The mushrooms at the end of the world*: on the possibility of life in capitalist ruins. Princeton: Princeton University Press.

von Schnitzler, Antina. (2016) *Democracy's Infrastructure: Techno-Politics and Protest after Apartheid*. Princeton, NJ: Princeton University Press.

Wark, McKenzie. (2017) "From Architecture to Kainotecture." e-flux Architecture, December 2017. https://www.e-flux.com/architecture/accumulation/122201/from-architecture-to-kainotecture

CHAPTER 5

Conflicts in the Resilient City

"What do you do when your world starts to fall apart?" (Tsing 2015: 01)

5.1 Ecological Cities and the Proposition of the "Green Common Good"

The spread of practices and responses to climate change, the emergence of new environmental groups and movements, the constant organisation of events with an ecological theme and the frequent protests demanding policies consistent with international climate commitments have provided fertile ground for alternatives to the Resilient City (RC) model to flourish. In this chapter, I contrast this institutional model with an emerging proposal formulated by activists: the idea of a green common good.

This common good should be understood in a broad sense, not restricted to urban green infrastructure. The data analysed here was collected through participant observation, social networks, articles and communiqués written by people engaged in the ecological transformation of urban territories. What emerges are conflicts between the official Resilient City and different proposals that I call Ecological Cities (ECs), which are impossible to abstract from the concrete contexts to which they belong.

D. Delatin Rodrigues, *Climate Change and Political Action in Italy*,
https://doi.org/10.1007/978-3-031-98771-7_5

The opposition is structured around the contrast between fossil fuel networks and infrastructures and low-carbon networks and alternative infrastructures. Those mobilising for ECs not only resists urban extractivism, but also seeks to create other ways of living, without resorting to urbanism as a privileged category. Local urban boundaries, in fact, "explode" in the face of the interdependent relationships they make explicit, connecting local actions to effects in distant territories. They point to the need to reinvent relational space beyond the human (Stépanoff 2024).

Returning to the land means relearning, affectively and technically, how to reposition oneself in the process of territorial transformation, making explicit the conditions under which modes of personal and collective reproduction are instituted—conditions that remain implicit in the daily normality sustained by the extraction of resources from other territories and bodies. The urban infrastructural apparatus, which guarantees circulation and access to goods, makes the impacts generated elsewhere invisible, maintaining local consumption based on non-human and unequally distributed functionalities.

It is therefore a minor policy that seeks to reverse the capture of territories by the imperative of growth. Personal and collective action cannot be dissociated from the semiotic, material and social flows in which it is inscribed (Deleuze and Guattari 1986). The interaction between climate activists generated new enunciations: narratives, slogans, meetings and demonstrations; new repertoires were activated from the "remains" of the modernist city: abandoned buildings, degraded areas and peripheries became privileged spaces for experimentation. Situational particularities of precariousness produced forms of belonging beyond the human (Puig de la Bellacasa 2017).

Although these responses are fragmented, their common trait is that they appear in the interstices (Stengers 2020) of the Anthropocene machine, sometimes in abandoned spaces, sometimes in small refuges that escape the extractive machinery. The chains of interdependence extend, generating new existential territories. The precariousness that runs through these practices is embedded in the landscape of modernity's failed promises: material abundance and ecological emancipation. The most difficult challenge is to make interdependent and multi-species relationships tangible in contexts still shaped by fossilised chains of dependency.

As Viveiros de Castro observes (2019: 296): "This is an age in which new ghosts, new monsters, new fears, new hopes haunt the imagination of what used to be called "the future" but is actually "the present", above all for peoples of all living species and other Terran entities who endure in a large part of the world now".

5.2 The Bassini Case

The Bassini case can be considered an emblematic event in the trajectory of the modernisation front. The decision by the Politecnico di Milano (POLIMI) to build a new building for the Chemistry department in an area previously occupied by a park was understood, institutionally, as part of regular urban development: revaluing an area, removing what already exists and erecting a new landscape.

But unlike the triumphant modernity of the twentieth century, this gesture was immediately questioned. Residents and students organised the Salviamo il Parco Bassini committee, which denounced the continuity of an urban planning model inherited from the 1970s—without transparency, based on the logic of concrete expansion (Mara, 15/12/2019). According to the activists, this was a mistake that compromised the principles of sustainability and revealed the institution's true alignment with property interests rather than environmental commitment.

Bassini, according to the accounts, was a green area dating from before 1722, never built on, and which maintained ecologically relevant soil. The conflict became a milestone in the opposition between fragments of urban nature and what became known as the Milan Model, orientated towards selling the city as an economic product, articulating growth and resilience (Sergio, student, 01/02/2020).

The defence of Bassini Park involved a broadening of the concept of the right to the city: not just human access to housing and services, but also the right to the existence and permanence of other forms of life. The committee not only defended the use of the space by students and residents, but also claimed the right of trees to reside in that territory.

In a communiqué dated 22/01/2020, the group stated its desire to discuss alternatives to the consolidation of the new building, on the basis that the park should be preserved. The question posed was: what kind of politics can emerge from this specific way of creating associations? (Kohn 2021: 71).

According to the committee:

> "Despite the felling of more than 50 trees, the ecological and social value of the area remains immense. That's why we're calling for real alternatives to rebuild the park. I don't believe that a free and transparent comparison is possible with the work in progress. Our request is for the immediate suspension of the works."

Urban Planning Secretary Pierfrancesco Maran proposed a working group with the Politecnico and the Università Statale to discuss a compensation plan, including a new green plan for the Città Studi neighbourhood. But the committee denounced its exclusion from the decision-making process:

> "The meeting didn't address our demands, nor were we involved from the start. We will take part, but we reaffirm that we don't want compensation, but a concrete alternative that doesn't destroy Bassini Park."

Since the Bassini case, new committees have begun to speak out in defence of other urban territories at risk. How can these areas be protected from capture by property capital? How can we prevent from being transformed into "dead lands" (Sassen 2014)? The case shows that the practices of destruction and reconstruction, far from being neutral, are forms of ecological and social dispossession.

In the Bassini case, various contentious repertoires were activated (Tilly 2008). This is a paradigmatic case for reflecting on the Milan Model: a city that tries to assert itself as resilient while reactivating the mechanisms that fuelled the ecological and climate crisis—those of accelerated modernisation and the Great Acceleration (Steffen et al. 2015).

The decision to eliminate Bassini was taken in October 2019, when "the professors of the Polytechnic and the residents of Città Studi abruptly woke up to the park being closed off by the construction site fences on 28 October", says the statement from the Save the Città Studi Committee.[1]

On 14 November 2019, an extraordinary conference was held at POLIMI, called by Rector Ferrucio Resta, to clarify the work on building the new Chemistry department. Faced with objections from various

[1] https://salviamocittastudi.wordpress.com/2019/11/23/qualcuno-sapeva-E-non-ha-detto-nulla/?fbclid=IwAR2u90SHjdgMwA05E2OFHTJcKFZMwND_4c3aRIddSKDtYmUJ_0V1xM1fLKM.

groups, the rector's response was that the project had followed normal bureaucratic procedures and therefore did not lack legitimacy. Residents, students and teachers questioned the transparency of the decision, since only the departments involved had been informed of the rector's plans. Events evolved quickly. Some attempts were made to block the decision to proceed, but it proved irreversible and the majority voted in favour of its continuation.[2]

On 10 December 2019, students and residents succeeded in blocking and suspending the work at Parque Bassini by occupying the construction site. Threats of resumption were met with new mobilisations and occupations of the site:

"This tree is mutilated, but it's still alive because the preliminary 'pruning' that was preparing to cut it down was stopped by the intervention of citizens. We must all mobilise, take part in the permanent vigil (from 7.30am to 5 pm), write to the institutions, put pressure on the Politecnico to review its project and save the park" (Comité Salviamo il Parco Bassini).

The construction blockade has become a common denominator of different struggles in Milan. More than a generalised tactic, it expresses the urgency of what they are trying to prevent: the destruction of urban territories under the pretext of modernisation. The effectiveness of these actions lies in the diffuse capacity to sabotage, by various means and on different scales: with a physical presence at the construction sites, with the broadcast in alternative media, with accusations of legal and technical non-compliance, with public demonstrations and institutional pressure (including members of the Green party taking a stand against the urban policy of the Sala administration, which they supported politically).

As Debinska (2021: 446) notes in a study on activism in Poland:

> "Climate change has acted as a common denominator for dispersed and localized environmental struggles, which have come to be framed as climate struggles. Local initiatives, such as the collectives that monitor and protest against the cutting down of natural forests and the destruction of nature reserves, have begun to emphasize the links between environmental destruction on the local level and global climate change. The climate apocalypse has become the horizon of almost any local environmental initiative. This has resulted in the growth and rearrangement of the

[2] https://www.facebook.com/ResilientGAP/posts/2690771974485878.

> assemblage of groups and people engaged in climate activism in Poland for noticing, monitoring and representing the effects of climate change on various planes: social, economic, environmental, political, psychological, etc."

The same dynamic can be observed in Milan. The counter-proposals formulated during the protests were based on a notion of interdependence, articulating local socio-ecological situations with the indeterminacy of the planetary climate system. Business-as-usual is not just economic continuity, but the destruction of precarious forms of interdependence.

Politecnico sustainability professor Arianna Azzellino (02/02/2020), also a member of the committee, said:

> "We need to consider urban green in its entirety - public, private, urban, peri-urban - recognising its intrinsic quality as a living system made up of animal and plant organisms, regulated by physical, chemical and biological flows that interact with the biosphere. This requires going beyond the concept of green as a qualitative urban standard and recognising its strategic value for integrated urban sustainability."

This vision, anchored in the natural sciences, also reveals a limit: the difficulty of formulating a political language capable of transforming this "natural exterior" into fertile soil for reimagining politics itself. While some activists operate with this techno-scientific grammar, others adopt more affective and relational registers: "The magnolia is leaving (…) We hope it survives in its new home, despite the cut roots, the relocation, the neglect. Insensitivity, obtuseness and human interests are tearing it from the land where it has grown in peace for so many years" (Save Bassini, 02/06/2020).

These two perspectives, although distinct, are not mutually exclusive. Both seek to include what modernist planning ignores. They establish another plane: that of experience. And they allow us to observe how subjects—human and non-human—are distributed in the problem scene.

As Azzellino writes again:

> "Planning parks and gardens is not just about providing a public service: it involves naturalistic, landscape and environmental values, as well as affective, aesthetic and practical meanings attributed by residents."

I can then return to the question formulated by Kohn among the Runa of Ecuador: What kind of politics can emerge from this peculiar way of creating associations? In an active world, there is no politics without non-human agentivity. If collective life is shaped by associations between humans and non-humans, then every decision about socio-material and ecological continuity also involves deciding who is included and who is not in this collective.

The mobilisations continued even between 25 December and 1 January: "On Friday 27 December, work on the destruction of Bassini Park can resume at 7.30am. At midday, lunch with Christmas leftovers. There will be live music. What to bring: camping tables and chairs, leftovers (sweet and savoury), drinks, plates and cutlery".

But on 2 January 2020, surprised by a strong police force, the activists were unable to stop the trees from being cut down:

> "This morning at 6am a large detachment of police escorted the workers who were to cut down the trees in Bassini park to the construction site. The activists present to guard the park were completely surrounded to prevent them from blocking the works again. At the moment, the workers are felling the park's century-old trees. Even the assurances of the municipality that had promised to obtain more in-depth assessments of the damage of the tree felling were useless."[3]

The Bassini case not only reveals the material and symbolic limits of the resilient city, but also the powers and tensions of an emerging politics: a concrete attempt to rebuild socio-ecological collectives based on the active participation of human and non-human agents in the constitution of the urban commons. On 2 January 2020, the physical existence of Parco Bassini came to an end. However, this outcome did not mark the end of the mobilisations. On the contrary: it was the very path of destruction—marked by the exclusion of citizens, repression and institutional insensitivity—that catalysed new articulations. As a result of the collective trauma generated by the elimination of the park, various city committees began to meet with the aim of discussing and combating what came to be known as the Modello Milano.

[3] The video explaining the event can be found here: https://www.facebook.com/permalink.php?story_fbid=1205422446316315&id=968811149977447&notif_id=1577960597973097&notif_t=page_post_reaction.

The Bassini case acted as a trigger event: it realised the real and concrete threat to the survival of other green areas in the city. For the activists, it was urgent to build a common front for the defence of the shared green, of all those who were "fighting for the pieces of green" in Milan. As D. expressed it, there was a clear need to "slow down the processes of urban development", which had been promoted under the rhetoric of resilience, but operated within the same logic of concrete expansion.

This attempt to coordinate struggles suffered a major setback with the outbreak of the Covid-19 health emergency at the end of February 2020. Several green areas were destroyed at a time when public demonstrations were banned and everyday presence in the city had drastically decreased due to teleworking and remote teaching.

P., an active member of the Resilient Gap collective, summarised the recurring cycle of instrumentalisation and delegitimisation that has marked the relationship between urban planning and popular protest:

> "The practice is always the same: projects are started and, once the fait accompli, new planting is promised (as if parks could be reduced to an arithmetic sum of trees). When protests arise, we are accused of being against the social interest of the new buildings. But Milan is Italy's second city in terms of built-up areas. Last year, 11 hectares of land were covered in concrete. Too much soil, too much cement" (P. 30/01/2020).

At the public assembly held on 2 February 2020, the different committees formulated their demands around two fundamental points:

> "That the Municipality of Milan bans the construction of new buildings in all green areas of the city; that a technical commission be set up with the participation of representatives from the green committees, responsible for assessing urban alternatives on a case-by-case basis that do not involve new land consumption."

This attempt to institutionalise an alternative urban planning practice marks the transition from a policy of resistance to a policy of proposition, anchored in new criteria of ecological justice, citizen participation and preservation of the commons. Here, the ecological city ceases to be a utopian abstraction and starts to be articulated as a horizon for concrete action in the face of the violence of extractivist urbanism. As noted by

Paolo Pileri[4] in an article on the site that sought to reflect what had happened to Bassini,

> "There have been emails, articles, discussions, petitions, assemblies, fortresses (I chaired them), marches to the mayor's windows to try to review the decision, albeit formally legitimate, taken a few years ago. But nothing could be done. In the early hours of 2 January 2020, protected by an unexpected cordon of local police, state police and *carabinieri*, the chainsaws went into action: 37 large trees were felled. The construction site could now begin".

5.3 After Bassini

But what happens after Bassini? The groups that were activated to prevent the destruction of the park began to consider that this event signified a particular turn in the environmentalism of the Sala administration, which was seeking to imprint its particular ecologist vision. According to the mayor, those who mobilised were the representatives of the "environmentalism of no", which could be translated as those who rejected Milan for big "events, real estate speculation and high property prices", as defined by T. (student, 27).

Paulo Burgio, reflecting on the political aspects of territorial planning, observes.

> "There is a real unwillingness to allow citizens to participate in social life, as is clear when Councillor Maran proposes a round table for citizens to define the means of compensation for the cutting down of the trees in Bassini park, a table to which a representative of the citizens' committee is also invited, when there is practically no reason for it, the trees have been cut down and that's that. (...) The main task and function of a councillor's office dedicated to responding to these requests continues to be to intervene to guarantee the recognition and implementation of the right to participate in issues relating to the city, the territory and the environment,

[4] https://altreconomia.it/perche-non-conviene-sacrificare-un-parco/?fbclid=IwAR0wvQLM4OlmE1cAjJTXq5weJC1p2RP7TW9IlvYdmqmEJxZzNZF7A4TSL0k.

> issues that are now crucial and unavoidable if we are concerned about the survival of the planet".[5]

The "Comitato salviamo città studi" has already outlined a difference that seems important to us. Using the Latin motto "*Ubi maior, minor cessat*", it explored the asymmetries as representative of a broader framework that included the Bassini case:

> "We *minors* don't understand why this construction site is in such a hurry to start up again. The world is at a standstill, we're all rethinking our lives and our priorities in the face of a pandemic, so why can't the Polytechnic (...) revise its plan to destroy Bassini Park and declare the death of the two cedars? I'm full of despair, but above all disappointment, with these *majors*."[6]

As a direct consequence of the destruction of Parco Bassini, the threat to other green areas in the city became more widely problematised. In response, an assembly of urban committees was called with the aim of articulating these struggles around the defence of the common green, as defined by the activists themselves: an attempt to bring together all those who "are fighting for pieces of green" in the city.

On 16 February 2020, representatives from Resilient Gap, FFF-MI, La Goccia, Baiamonti Verde Comune, Comitati Cittadini per il Verde, Climate Reality Project, Off Topic, Core Lab and FFF-Monza took part in a meeting to discuss how to contain the advance of Modello Milano. The proposal to create a new coordination generated hesitation, with fears about the proliferation of structures with little practical effectiveness. But the focus of the discussion was different: how to organise different perspectives and practices to resist the dynamics of urban occupation led by real estate speculation.

Reports like M.'s denounced the lack of openness on the part of the city council to alternatives to the Resistance Museum project, whose construction threatened one of the few green areas in Piazzale Baiamonti.

[5] http://www.z3xmi.it/pagina.phtml?_id_articolo=13600-Cittadini-E-Partecipazione.html&fbclid=IwAR12BiXXd4shH-T-xw8eDaI6Yd3tA-KFsMDzKW73o0Wl8i9PLcpOFztlNZc.

[6] https://salviamocittastudi.wordpress.com/2020/04/21/ubi-maior/?fbclid=IwAR2pSuqPC6Ghe75Ml-5x9c87eBTq-TZCguk4iy9lUsFP1xY-gZBbqZgB_y0.

R. emphasised: "We keep asking for a garden to be built instead of the museum. That's an area that should be preserved".

The group recognised that it was necessary to expand the public concerned about these issues, especially in view of the municipal elections scheduled for September 2021. Some proposals were aimed precisely at this goal: green benches installed in symbolic places; an expanded presence on social media and radio programmes; initiatives to promote the affective use of threatened spaces. As G. said, it was necessary to "make people live experiences that give a different meaning to places". F., from Resilient Gap, added that it was essential to carry out "actions with a strong communicative impact, capable of generating effective social engagement".

The type of engagement proposed—making diverse experiences in living places—indicates a repositioning of activism: from protest against destruction to care for permanence. As Haraway (2016: 31) reminds us: "No one lives everywhere; everyone lives somewhere. Nothing is connected to everything; everything is connected to something". Activists tried to make these connections visible through symbolic practices and staging that represented the urban climate crisis.

One example was the performance parade organised by the Collective Off Topic, with swimsuits in the Canal and a 400-metre banner. Another was the salamander parade, through which activists staged the resistance of endangered species to urbanisation. D., from the Climate Reality Project, suggested an action inspired by Extinction Rebellion: planting trees in front of Parliament against a pro-coal decree.

M., for his part, insisted on the need to challenge the argument that it is necessary to extract land resources in order to manage the city. J., from the FFF-MI, emphasised the importance of networking with the green committees to build "a truly sustainable and resilient city", and announced that the 5th climate strike would include a night demonstration. This demonstration, however, was cancelled with the arrival of the Covid-19 pandemic.

Co-operation between the committees continued, despite the evictions that followed. On 8 July 2020, the Baiamonti area was closed again to begin construction of the Resistance Museum. The Baiamonti Verde Comune collective published a long note denouncing the destruction of the community garden:

> "Why did they destroy the amphitheatre and the plants we've been looking after for a year? Why did they fence and lock the area? Is it a crime to look after an abandoned green area? We created a space where people could meet, watch films, have picnics and rehearse. And now they've destroyed everything".

Despite this, the space was reoccupied in February 2021. An agreement was even negotiated with the town hall to allow cultural activities until December 2022. However, just five days later, bulldozers returned to the site without warning. In a statement, the activists declared:

> "They mocked us. We found out that they were starting to build the second pyramid, which was decided in September 2020. We can't stop the bulldozers, but we can tell the citizens what's going on".

The image of the "death of the park" reappears: the uprooting of the trees and the people who looked after them is narrated as a double ecological and social violence. Azzellino then proposed that the project be re-evaluated on the basis of "ecosystem services". The aim was to use the very instruments of urban management to halt the logic of destruction. However, this strategy also highlights a risk: in an environment saturated by quantitative rationality, even attempts to break away are easily absorbed into the administrative language of the market.

The transformation of the city thus becomes a field of practical dissociation. The subjects are caught between the desire to stop business-as-usual and their daily dependence on the extractive infrastructure. This dissociation is practical: it concerns the difficulty of transforming the material conditions of existence.

In this context, anti-political rhetoric, with its promise of total change to the system, can sound attractive. But in practice, what is immediately problematised are the local, everyday, relational elements of urban life. According to ISPRA data, between 2006 and 2020, more than 2,153 hectares of soil were consumed in the Milan Metropolitan Area. A figure that sums up the pace of extractive urbanisation—and at the same time indicates the fragmented field where dissent and creation meet.[7]

[7] https://www.isprambiente.gov.it/it/istituto-informa/comunicati-stampa/anno-2021/neanche-la-pandemia-ferma-il-consumo-di-suolo-speciale-roma-E-milano.

According to Michele Munafò, ISPRA's rapporteur on soil consumption in Italy.

> "In addition, the current soil situation must be considered, with 58% of the municipal area, or 10,500 hectares, now artificialised: a territory in a clearly critical situation. Rome - to make a comparison - has consumed 23 per cent of its area, although built-up areas cover a much larger area, 30,000 hectares".[8]

Among the urban projects that most threaten the Municipality of Milan are the new San Siro stadium, Goccia Park and the Seveso rolling mill. For the first, 5 hectares are to be cut down; for the second, 40 hectares of spontaneous vegetation are to be destroyed; for the third, 4 hectares are to be cut down. Resistance to the expansion of the extractive town is coloured by a melancholic tone: business continues as usual, immune to the ongoing interconnected crises. A sense of inevitability permeates the experience of those who have become active.

For the Save Bassini Committee, "the institutions have not yet realised that this type of development is unsustainable. The symbol of our protest as conscious citizens will be the two remaining cedars, mutilated but still alive" (13/02/2020). With the emergence of Covid-19, the continued destruction of green areas has become even more absurd. According to the activists, the pandemic should have imposed "a review of our development model; only this time, it's not just environmentalists who are saying this, but also doctors" (04/03/2020).

The understanding that connects pandemic, climate crisis and territorial destruction has become a common point of convergence: "It is absurd to believe that, despite pollution, pandemics and the climate emergency, we can continue to consume soil, build and cut down trees. Bassini Park was more than an urban forest: it was a small piece of the future we were betting on" (01/02/2021).

In Goccia Park, the local committee insisted that "Goccia is an opportunity to change the point of view on the city project. We need to build participatory transformation from within the area, with the real involvement of the inhabitants" (02/10/2021). In the case of San Siro, the

[8] https://www.firstonline.info/consumo-del-suolo-munafo-ispra-milano-non-E-tutto-oro/?fbclid=IwAR2lq4ODRbBkql97EgtNeyldTkcPXJdTx0Pz7ID-Q_R3lhhJ1erenZAYK2M.

activists denounced the replacement of the stadium with more concrete: "It is possible to renovate the Meazza. The project has been presented. The future has changed, Milan's planning must change too" (25/01/2021).

Criticism converged around the same question: how can we promise an ecological turnaround and at the same time continue to destroy the rare green spaces that exist? As the Baiamonti Verde Comune collective pointed out (26/02/2021): "Sala promises a serious environmental turnaround. The Greens believe in it and are going with him to the elections. But in the meantime, a pact with citizens is being broken. A new garden is about to be sacrificed to make way for a second pyramid of glass and concrete in the same square".

In the end, Giuseppe Sala was re-elected in September 2021, now as a member of the European Greens in a broad political coalition. One could conclude this chapter with the realisation of a clear defeat for activists in the face of the urban machine. But I prefer, despite all the evidence to the contrary, to insist on what resists this defeat.

Defeats are not definitive closures. They reveal areas of openness. They show that elements considered irrelevant by managers and planners can emerge as central to other forms of collective life. The mutilated existence of a cedar tree, the performative protest in front of a construction site, the precarious permanence of an urban garden are at the centre of another way of thinking about the urban: not as a function, but as a relationship; not as a plan, but as a process of shared inhabitation, necessarily situated and multi-species.

References

Debinska, Maria Salomea. (2021) Witnessing from within: hyperobjects and climate activism in Poland. *HAU: Journal of Ethnographic Theory*; 11(2): 445–460.

Haraway, Donna. (2016) *Staying with the trouble*: making kin in the Chthulucene. Durham: Duke University Press.

Kohn, Eduardo. (2021) Come pensano le foreste: Per un'antropologia oltre l'umano. Translated by Alessandro Lucera and Alessandro Palmieri. Preface by Emanuele Coccia. Milano: Nottetempo.

Puig de la Bellacasa, Maria. (2017) *Matters of care*: speculative ethics in more than human worlds. Minneapolis: University of Minnesota Press.

Sassen, Saskia. (2014) *Expulsions*: brutality and complexity in the global economy. Cambridge: Harvard University Press.

Steffen, Will; Broadgate, Wendy; Deutsch, Lisa; Gaffney, Owen and Ludwig, Cornelia. (2015) The trajectory of the Anthropocene: the Great Acceleration. *The Anthropocene Review* Vol. 2(1) 81–98.

Steffen, Will; Katherine Richardson; Johan Rockström; Sarah E. Cornell; Ingo Fetzer; Elena M. Bennett; Reinette Biggs; Wim de Vries; et al. (2015) "Planetary Boundaries: Guiding Human Development on a Changing Planet." *Science* 347 (6223): 1259855. https://doi.org/10.1126/science.1259855

Stengers, Isabelle. (2020a) *Réactiver le sens commun*: lecture de Whitehead en temps de débâcle. Paris: Les empêcheurs de penser en rond.

Stépanoff, Charles. (2024) Attachements. Enquête sur nos liens au-delà de l'humain. Paris: La Découverte.

Tilly, Charles. (2008) Contentious Performances. Cambridge: Cambridge University Press.

Tsing, Anna. (2015) *The mushrooms at the end of the world*: on the possibility of life in capitalist ruins. Princeton: Princeton University Press.

Viveiros de Castro, Eduardo. (2019) On Models and Examples Engineers and Bricoleurs in the

CHAPTER 6

What Should Be Done?

"Much has already been written about the Anthropocene—and we have barely started—yet one wonders how little of consequence has been said about "what is to be done" (and undone), not to mention the little that is being done." (Viveiros de Castro, 2019, S296)

"N. reports on his experience with Milan for the climate so far: it brings together more or less radical realities, it's difficult, but it's the only way not to be insignificant. According to V., Milan for the climate was born because the various realities realised that, on their own, they don't weigh enough, so a common denominator has to be found. D. recounts the birth of Milan for the climate: the aim was to tackle the climate crisis. It was known that the various realities had different approaches, but it was necessary to come together, and diversity can be a resource. According to I., we have to give ourselves rules because, ideally, diversity is wealth, but in reality it has created a lot of problems, and some people don't come here anymore". (02/02/2020—M4C meeting document)

6.1 Climate Activism and the Problems of Political Action

In December 2018, I began to observe what would become one of my fields of research: the subjects who became active. After the start of Greta Thunberg's Friday strikes, also in Milan, the movement was beginning

D. Delatin Rodrigues, *Climate Change and Political Action in Italy*,
https://doi.org/10.1007/978-3-031-98771-7_6

on the initiative of an American citizen with 30 years' residence in Italy called Sarah Marder:

> "Whoever you are, wherever you are, we need you to stand up now in front of parliament or your local government office to let them know that we demand climate action." These words, spoken by Greta Thunberg on 12 December 2018, changed my life. Greta is the sixteen-year-old Swedish activist who has made headlines and won the hearts of people of all ages around the world who are fed up with the lack of climate action from world leaders."[1]

At first, as several newspapers and the activists themselves acknowledged, the people who gathered in Piazza Scala—the seat of the municipal administration next to Milan's cathedral in the city centre—were few in number:

> "When I heard Greta Thunberg asking for everyone's help to take part in this battle, I decided to try to organise a climate strike in Milan too", explains Sarah. The first time I went on strike, following Greta's example, was on 14 December last year. At first, I was alone. The following Friday there were six of us, the following Friday there were twelve, and so on. Many cities are mobilising now, there's a lot of enthusiasm everywhere."[2]

On 15 March 2019, two massive demonstrations swept through the centre of Milan. More than 100,000 people took part in the morning event organised by the Milanese branch of Fridays for Future (FFF), followed by a second demonstration with around 40,000 people. It was the largest climate mobilisation ever held in Italy, generating tangible euphoria among activists. Many believed that this moment represented a historic turning point. At the end of the afternoon march, an improvised loudspeaker attached to a bicycle announced a future assembly to discuss how Milanese groups engaged in the climate crisis could articulate a common action, both on a local and wider scale. This call was the starting point for the creation of the Milano per il Clima (M4C) network.

[1] https://www.pressenza.com/2019/01/environmental-optimism-over-despair-cities-are-key-to-fighting-climate-change/?fbclid=IwAR3I9UERupVHa_suBY4QTba7Sq_1qv96yWa2_eWUwRoMeoDvmp1-E3BN8UI

[2] https://www.ohga.it/climate-strike-A-milano-grazie-A-fridaysforfuture-si-sciopera-contro-il-cambiamento-climatico/

The first news about the new network circulated on social media forums and alternative communication channels. In April 2019, I signed up to M4C's private Facebook group and asked to join the WhatsApp group, introducing myself as a PhD student researching climate activism in the city.

6.2 Digital Dynamics and Articulation Practices

The material that makes up this chapter derives from participant observation at events, ethnographic notes and the use of digital platforms, including Facebook and WhatsApp groups and the network's repository of shared documents (Google Drive). Although these sources are limited to the discursive plane, I choose to treat them as records of the attempt to transform fossil urban socio-materiality. I seek to understand how activists make their difficulties explicit, negotiate alliances and try to produce broader political effects. In other words, how they endeavour to "contaminate" the urban environment with new forms of relationship, removing it from the chains of dependency that sustain fossil socialities.

The platforms performed different functions: Facebook was a space for information and public debate, even though it was constantly crossed by disputes over content control; WhatsApp functioned as a fast channel for internal communication and mobilisation, although it generated discussions about the degree of autonomy of the groups and the limits of the coalition. This chapter focuses especially on the documents shared (meeting minutes and manifestos), through which it is possible to identify which organisations participated, what topics were discussed, what criticisms and proposals emerged, and how the prospects for action were drawn up.

Between May 2019 and September 2020, the network organised 24 formal meetings. Joining M4C was simple: you just had to answer the call on the Facebook page to be added to the communication groups. The WhatsApp group, with around 60–70 active members, was not the privileged space for decision-making, which took place in face-to-face meetings.

Despite the large number of participants, the coordination meetings were attended on average by representatives of between 5 and 15 organisations. Some attended with more than one member; many activists participated in several organisations simultaneously, as was the case with students from the Tempo Zero collective (formed by physics students from the Università Statale) who were also part of the FFF-MI. This overlap expresses the reticular nature of climate activism and the centrality of situations in defining modes of engagement.

Each group brought with it its own repertoire of practices and interpretations of the climate crisis. The confluence around the M4C did not imply homogeneity, but openness to experimenting with multiple forms of articulation. This became clear at the FFF-Italy national assembly held in Milan on 13 and 14 April 2019, shortly after the historic demonstration in March. It was on this occasion that the first debates emerged on how to situate climate change in concrete urban territories and how to connect local struggles to horizons of systemic transformation.

From here on, the chapter will continue to explore these dynamics, following in a denser way the practical and discursive compositions that emerge from the assemblies, the impasses faced and the gestures of reimagining the urban as a field of climatic struggles.

6.3 "The Time Has Come; Now What?"

At the first national assembly of Fridays for Future-Italy, held in April 2019, I was able to observe how activists from all over the country sought to justify their commitment and outline possible paths for collective action. One of the central axes of discussion was the relationship between science and politics. In various interventions, participants sought to contrast the robustness of scientific data with the passivity of the institutional response. Representatives from Genoa, Monza, Piacenza and Val di Susa emphasised, each in their own way, the centrality of science as a basis for policy and the urgency of "following the data".

This insistence on transforming politics through the authority of scientific evidence will repeatedly run through our ethnography. Data is mobilised to put pressure on decision-makers, but also to shape public opinion, to transform collective habits and to justify performances that dramatise planetary collapse and, at the same time, exalt human ingenuity. Scientific communication itself becomes a field of aesthetic and militant invention.

But what the activists also emphasised was the need to protect this camp from partisan capture. Representatives from Campobasso, Siracusa and Molise insisted that the movement should remain anti-fascist, non-partisan and horizontally structured. Mobilisation, they said, should be bottom-up, anchored in the territories and expressed in open assemblies. Activist practice should be itinerant (Bolzano), support local causes (Vittorio Veneto) and become the "voice of the territories" (Campobasso/Molise).

The link between science and politics is a central issue in current climate activism, but it's not new. Science and technology studies have already shown the challenges of directly transposing laboratory facts into collective life. In the Milanese context, scientific data—especially IPCC reports such as the 2018 report on the 1.5°C limit—is not deconstructed. On the contrary, they are treated as essential components in the training of citizens and climate activists.

The discussion about the relationship between scientific knowledge and responsive politics also cuts across an important theoretical debate: the notion of post-politics. This approach argues that the fight against climate change has been structured in such a way as to displace politics in its own sense, replacing it with technocratic management and consensus building. As Swyngedouw (2010: 215) puts it:

> "This post-political frame is structured around the perceived inevitability of capitalism and a market economy as the basic organizational structure of the social and economic order, for which there is no alternative. The corresponding mode of governmentality is structured around dialogical forms of consensus formation, technocratic management and problem-focused governance, sustained by populist discursive regimes."

Although I consider this critical approach to be relevant—especially to highlight the risks of climate agendas being captured by market logics—I disagree with its use as a totalising diagnosis. By adopting this model comprehensively, we run the risk of invisibilising practices of dissent, invention and political reconfiguration that take place on the margins of (or against) conventional institutional spaces.

Activist politics is not just about contesting existing institutions, because its demands also imply the creation of new spaces and forms of decision-making. The real question, then, is not just whether science becomes an instrument of technocracy, but how it participates in the formation of publics that face problematic situations in which the existing political system proves insufficient (Dewey, 2016).

My aim is therefore to accompany these experiments and rescue them from the systematic disqualification that comes both from above (from institutional elites) and from critics who, in the name of anti-systemic purity, refuse any form of partial engagement. No positive description of terrestrial politics (Latour, 2018) will be possible without following these collectives in their ability to create situations, set limits and experiment with ways of intervening in the conditions of collective life.

6.4 Become a Climate Activist

Those who identify themselves as climate activists act on multiple fronts, even entering areas considered legitimately "political"—the same areas demarcated by institutional representatives such as the MEP from the Democratic Party, who contrasts youthful impetus with the responsibility of formal bodies. This contrast between ways of doing politics is productive for understanding not only what counts for those who become active, but also how the materiality of everyday life is constantly invested with political relevance.

As Knox (2020: 213) notes,

> "When it comes to climate change, critics are right in pointing out that climate activists are deeply entangled with technocracy. However, approaching their actions from the perspective of thinking like a climate rather than from a dematerialised view of political thought creates the possibility of seeing that this entanglement is not about a narrowing of politics but rather an opening to a version of the political that is capable of incorporating the communicative capacities of nonhuman and human forms by attending to the way they are described by science."

In this sense, political practices are not separate from socio-material and energy transformations. It's about how we move, what we consume, what energy sources would be appropriate to reduce emissions, and how to transform the infrastructures that sustain social and economic life. However, recognising these problems is not enough: we need to discuss how these issues can be articulated in transformative collective actions.

Activists become active through situated experiences rooted in specific environments. As Stengers (2020a, b: 78) writes, "ce sont les faits obstinés, capables de faire obstruction, interrompre impitoyablement, la marche d'un raisonnement ". These local conditions not only prevent certain courses of action but also favour others. It is therefore inappropriate to think of subjects as detached from their environments. On the contrary, subjects emerge from the material and symbolic relationships they have with their environment. The environment is therefore a constitutive dimension of political action.

The problematic situations that activists face in their daily lives challenge them: What to do in the face of the ongoing catastrophe? Without ready-made answers, they must position themselves in the face of this injunction. "Becoming active" is a partial experience, made up of personal and collective, material and symbolic ties. But it is not a passive reflection of the environment. On the contrary, it is a conceptual process: activists elaborate their experience, build diagnoses, formulate hypotheses, generate intelligibility about the conditions that limit or enable their practices.

This process also involves analysing structuring forces. Concepts such as "chains of dependence" are mobilised to explain why certain practices are reproducible and others are not. They assess institutional capacity and will to transform climatic conditions. Local actions are thought of in relation to extra-local chains. Infrastructures, ecologies and territories are reconfigured through attention to the links that articulate them.

Among the student collectives, committees and associations in the Milano per il Clima network, the scope of action is diverse. Annex 1 summarises the areas in which these groups operate. I don't intend to give a complete history of each of them here, but rather to identify how activists make sense of their practice and how they articulate their "active" status with ecological and climate demands.

Based on the minutes of the meetings, field notes and discussions in the M4C WhatsApp group, it will be possible to extract the most relevant axes that guided the debates and the difficulties of political articulation between the groups. The question can be formulated as follows: How can the multiplicity of subjects affected be transformed into a common political force? To this end, I have chosen to quote excerpts from the documents and conversations in full, in order to offer a localised description, in the words of the actors themselves. The tactics and conflicts surrounding these attempts at articulation will be explored in the next chapter.

6.5 Where to Land?

"Connaître, c'est toujours connaître quelque chose qui endure et tant que capable d'endurer. Mais connaître un être, c'est être situé par lui, par la manière dont il valorise son environnement dont fait partie celle qui cherche à connaître - c'est apprendre quelles questions cette valorisation rend pertinentes" (Stengers, 2020a: 97).

How do we situate climate change? And even more so, how do we situate ourselves personally and collectively in the face of this event? What consequences does this problematisation have for the constitution of collective coping practices, whether in the form of anti-politics or the alter-politics proposed by Ghassan Hage (2012)?

My research strategy consisted of following what mobilised the activists, what made them think and act together to create an uncertain collective. As we will see, they themselves were hesitant to define the nature of the group that emerged in 2019: would it be a network, a movement, a platform? This indeterminacy stems from pragmatic questions about how to respond collectively to the climate crisis, how to expand their actions to regional, national and international scales, and how to imagine ways of life organised less by chains of dependence and more by relationships of interdependence.

In one of the first documents shared on the M4C WhatsApp group, the following question appeared: "Local–global. How do we position ourselves in relation to the actors of the territory and what vision do we want to give in relation to the global context?". And also: we need to "strengthen, as a citizen network, the value of science in the fight against climate change" (meeting of 07/08/2019).

The M4C network was made up of a heterogeneous constellation of people and organisations that were already active on various fronts linked to the environment in the broadest sense, and which were called upon to reformulate their practices in the light of the climate emergency. But there was no ready-made answer on how to do this. The challenge was precisely to articulate different responses, coming from different areas of social life and different practices.

A response produced in a specialised field may be effective in that context, but there is no guarantee that it will work in others. As Stengers (2015) argues, an innovation tested in an "experimental environment" will not behave in the same way when transposed to other environments.

Several groups that were active in the M4C coalition were recent formations, emerging in the wake of the "Greta Thunberg event". They exemplify the emerging dimension of public and climate politics: not only as anti-politics, but also as experimental alter-politics, with all the risks of capture by technocratic post-political management. These collectives are organised based on the territories in which they exist, forming alliances that allow them to expand their spaces for action and experimentation.

6.6 Activation Paths

In their first year, these activists tried out different forms of activation, their own or those of other agents. They took part in meetings with municipal and regional administrations, met political figures, established contact with companies and other associations; they tested different formats, such as the typical university conferences, participation in other associations' events; they organised two climate festivals[3] with different institutions; they produced a lot of material on the climate crisis and various proposals on how to respond to it. Reports, analyses, discussions, banners and posters; demonstrations in squares, large parades, online protests, on bicycles; direct actions and occupations, theatrical and musical performances, exhibitions. They sought to infect the environment to create a suitable terrain for the low-carbon transformations they considered necessary. These events must be considered in terms of the specificities of their practices and the public they were aimed at. I cannot judge whether they are valid or not. But they can all be defined as practices that seek to break the consensus and "attachments" that guarantee business as usual. For Simpson (2020: 17).

> "The ensemble of ways that urban space is made and remade to facilitate and conform with the dictates and accumulative logics of fossil fuel industries. Fossil fuel extraction and consumption have given rise to carbon cities that are underpinned by (and extend) forms of racial, colonial, capitalist, gendered, and environmental violence and injustice. These exploitative and harmful relations are the conditions of possibility which enable the production and reproduction of urban space."

They are trying to become a reliable interlocutor with the city administration to intervene, lobby and help, according to them, in the "elaboration of a new vision of the city" (F.). This process can be considered

[3] Which at the end of 2021 became an association. https://www.klimatfest.org/2021/.

as far as the "Bassini Affair", which will provoke a change in perception regarding the city administration's commitment to undo what Simpon (2020) calls fossil urbanism and move towards a form that could broaden the aspects considered relevant for the planning and transformation of cities.[4] The difference between climate urbanism and fossil urbanism is that if the former provides the analytical tools to understand the ongoing socio-material transformations of urban interventions on climate change, the latter provides critical instances to apprehend it as an artefact that resists phasing out transformations.

The difficulties of being heard or considered as relevant voices in the public debate on climate change have led these activists to seek some kind of dialogue with the city administration. But how can they find a way to be heard? And, perhaps most importantly, with what language to speak to be heard? For them, this meant getting inside the administration's documents and plans and learning its managerial and technical language. In this sense, they began (10/04/2019) to think about actions that could be taken to establish a relationship with these institutions as a competent subject: the first step would be to study the Sustainable Energy Action Plan (Piano di Azione per l'Energia Sostenibile (PAES)); another was the idea of inviting experts to hold training seminars for the group's members on the subject:

> "Propose a series of questions to be asked of the municipality (for example, how will decarbonisation be carried out, why have we only reduced by 11% when the target is 20%, how to improve communication with the PAES; 4. To have a relationship with citizens, the idea could be to make a path of deliberative assemblies; on this, the process should be designed and also a test phase of it".

Among the proposals made at one of the meetings (20/10/2019), three emerged: "Interview the institutions" (**P.** Resilient Gap); write "an open letter to Mayor Sala by 20 November", together with the FFF (**J.** Tempo Zero); and the proposal that M4C's thematic groups on technique and relations with institutions could work together (**D.** Climate Reality).

The relationship with political agents, the need to think of an expanded action that could include other subjects outside the associative or student

[4] In May 2019, the city council approved a climate emergency declaration.

realities was the reason that led the activists to organise two demonstrations during the climate march—an atypical situation worldwide, according to them—instead of just one; the aim was to create "opportunities for workers and link two different categories: students and workers, citizenship in general" (24/07/2019—documents). This point brings us back to the observations made by Sarah Marder at the start of the climate demonstrations. Most of the people who gathered at the Friday strikes in front of City Hall were not exactly young. During all the demonstrations, it was clear that the morning one was organised by the FFF and the evening one by the M4C. But even at the FFF's open assemblies, the presence of people over 45 was evident.

One of the initial questions asked of the activists—that is, the ability to activate and attract other people to the demonstrations and how this could contribute to taking a public stance on the climate situation—was extremely relevant:

> "He emphasises the importance of involving small territorial associations and also of re-evaluating traditional forms of involvement (banquets, gazebos, etc.). It's parallel work of equal importance to the institutional level. Put down roots in the territory, raise awareness, broaden the base. The ability of the people who take part in the banquets to engage and communicate complex issues in a simple way is necessary, which is why we propose training sessions for these people, focussed on exactly these skills" (24/07/2019—documents).

However, bringing heterogeneous realities into relationships has brought practical difficulties in articulating a shared common perspective.

> "We are a network of associations that have no experience on the streets, apart from the students, and the perception is that there is no associative line. A manifesto has been proposed. Is there a will to do this? Yes and no. The general feeling is that we're chasing events and we're only working to organise something together with the FFF. We're an "event machine". Perhaps M4C is going too far to become an association rather than a network?" (07/10/2019—documents).

To solve this problem, they propose moments of "political confrontation, structuring, dates are necessary. Course to become a facilitator (meeting with facilitator) self-training/political assembly/public event. Deliberative assembly" (07/10/2019—documents). This is an important point; on several occasions, the question of what kind of organisation should be formed was put on the agenda; the network sought to articulate diverse subjects, with heterogeneous organisational characteristics, origins and modes of action (formal and informal groups, professional activists). Perhaps it would be more appropriate to say that it was looking for ways to articulate several diverse networks into a single network—some more traditional NGO networks, others newer, such as the new climate collectives. Articulation would—presumably—make it possible to unify efforts and political power. Their task would not only be to force policymakers to act, but also to propose some solutions to the problems arising from climate change.[5]

To do this, it was necessary to take advantage of important events or dates in the city's calendar to introduce a counterpoint to the debates on climate change and not limit themselves to simple, routine mass demonstrations. Despite the large demonstration in the first year, it became clear to the activists that this was not enough. To combat business-as-usual, they needed to put down roots in the material fabric of the territory, but also in the imagination of its inhabitants; to combat it, they needed a presence that offered new possibilities for living together. One of them was Black Friday. As reported in the Milanese section of the newspaper "La Repubblica" on 29/11/2019, young activists gathered in the city centre to protest against what they defined as a date of consumerist celebration:

> "Hence the young people's decision to use today to protest against "consumerism and overproduction, which are also the causes of the current environmental situation", explains another organiser, Alessandro Silvello. We decided to call the day 'Block Friday' precisely for this reason, to highlight our protest and draw everyone's attention to the effects of consumerism taken to the extreme, of which Black Friday is a symbol: I'm thinking of the reduction in rights and wages of many workers employed in this system."[6]

[5] This didn't exclude attempts to "control" the organisational and political line of the network; there were constant conflicts, many of which revolved around the use of social media for the publication of events, the publication of activist actions in Europe and articles and, above all, about the methods deemed necessary for business as usual (creating a clear division between those who advocated direct action and those who didn't).

[6] https://milano.repubblica.it/cronaca/2019/11/29/news/fridays_for_future_black_friday_milano-242190222/.

This "consumption date", like other dates (Christmas, Easter), was chosen to produce an inversion of their meanings. I could even say that this is a process analogous to the inversion of infrastructure, but in this case, the political objective is to make explicit the chains of dependencies—and their consequences—necessary to produce this moment of collective consumption and its desires in everyday life. Other dates are chosen as opportunities to use the public meeting to "hold the city administration to account". But even this gesture produced divisions between the groups that sought to avoid direct confrontation with the public authorities: "During the last meeting, the possibility of organising a public assembly during Climate Week with Milan Climate Action and institutions was discussed, in order to confront the administration about what is being done and what has been done for the climate emergency".

According to what was recorded in the documents, regarding the possible relationship between the activist and the local administration, two "contrasting opinions" were presented. The first was that "some would not agree because this could offer an easy side for the media use of the initiative and also because there are already elements for the administration to act on and it only needs to do so"; the second was that "others are of the idea of building a moment of confrontation in which the associations and the administration meet face to face to express requests and possibly proposals". The finalised decision "was forwarded to the next meetings".

When I observe that becoming active "activates" a process of explicitness in which "everything" must be considered, we will see that activists begin to collectively ask themselves questions whose unresolved resolution requires the intersection of different spheres of action. They oscillate between autonomy and commitment, between critical political engagement and the risks of being "instrumentalised" and co-opted.

These experiences include everything that makes up the subjects' experiential field and not something that is exclusively economic or exclusively climatic. In this sense, they must "se familiariser avec, pister, enquêter, essayer, induire per analogie, découvrir—comprendre peut-être, mais jamais en general, toujours relativement à une question, une préoccupation ou une situation" (Stengers, 2020a, b:79).

We could even say that this is a criticism of the limits set by specialists—be they economists or engineers—who, by circumscribing an aspect considered relevant to their own practice, seek to impose themselves as general assessment criteria. In this sense, they pointed out that it was necessary to "understand the concept of development" (**P.** RG 20/10/2019), as well as to understand how to "push politics to really engage" (**F.** RG, 20/10/2019). But it was also posed as an injunction in which it was necessary to produce a minimum consensus through which actions could be organised. According to **M.**, this could be done by structuring a programme document based on the limits pointed out by the IPCC report published in 2018, which establishes 1.5 °C as the limit temperature that the Earth's climate system could withstand without "entering completely uncharted territory". As we saw at the FFF national assembly, the issue of the 1.5 °C temperature increase limit was also a limit to the possibility of collective action: it was the lowest common denominator that could bring such different realities to an agreement:

> "I like the structure of the "1.5°C target" point. Add a chapter on what we want to do. We need this work above all else; **G.**: "So what do we want to do?" is missing. He proposes structuring the document in "our values, our requests, our method"; **J.**: adds a reference to migration; **L.**: suggests not stopping at the pillars (the basics), but going ahead and asking what the objective of this work is; **S.**: suggests that we define which areas we want to focus on: just energy and mobility, or more?"

Activists are drawn into other spheres from the moment they start asking questions about the links between consumption and emissions. It seems to us that this marks a difference that inevitably leads them to develop a political perspective that cannot be limited solely to a confrontation with the system—along the lines described by Hage (2012) as anti-politics.

These doubts are therefore reflected in the definition of the problem they must tackle (with its multiple connections) and the tactics that could be pertinent to disrupting business as usual. As we will see in the next chapter, the decision to adopt direct action tactics is rejected by part of the group, especially by an activist linked to The Climate Project Reality (an international NGO founded by former US Vice President Al Gore), who a priori rejected any direct political action; the conflictual repertoire activated by the activists was seen as misguided and, above all, dangerous—which is why it is constantly associated with the antagonistic practices of the Italian extra-parliamentary left of the 1970–80 period (Balestrini and Moroni, 1988). The defence of absolute pacifism (Malm, 2021) delimits the modalities of action as a practice of persuasion and dialogue—which, according to some of them, would increase public support to push even harder for an energy transition.

In this case, there was a deep divide between a strictly institutional line and another that sought to "contaminate" all aspects associated with the urban fossil machine, with the values associated with it, and that saw the interruption of normal daily flows as an opportunity to affect business as usual, as well as the consensus that supports it. Faced with a proposal to carry out a traffic blockade, the Climate Reality activist articulated his rejection, describing it as ineffective and in which he risked losing existing public support; according to him (10/04/2019):

> "As I've already said, I'm against blockades; in fact, I consider them counterproductive: the only 'advantage' I see is that they are favourable to the media; for the rest, we make the majority of the population enemies, and the institutions won't want to listen to us or talk to us. So if we want our pillars to be fulfilled, from my point of view, blockades are counterproductive. I call for an official vote on what to do: blockades or no blockades? For me, this is a decisive point, so I need to know what has been decided. Organising (a demonstration) together in the morning is not feasible without authorisation, as it is incompatible with the code of conduct of many associations."

In the end, the decision to participate was rejected and, according to **J.** (Tempo Zero/FFF-MI), it was "understandable that if the associations are not close to conflictive practices, people can participate individually". The tactic of blocking and direct action was proposed by the student collectives; received with silence from the majority, or with messages of support or disapproval, it meant interrupting the flows (communicative, administrative) in a way that represented a crisis and made explicit a critical dimension of business as usual. It was common to hear the slogan at demonstrations: if nothing was done, the protesters would block the city—something that never happened. In conversation with some activists, it seemed clear that blocking the city would be technically feasible, as it would be enough to "block a few nerve centres in the city for traffic to collapse" (**D.**, 29-year-old student, personal communication—08/02/2020). Even if the position criticising this tactic shouldn't be considered only from the point of view of the individuals opposed to the bloc, in fact this division is not exclusive to Milanese activists—as demonstrated, for example, by Malm (2021).[7] Explicit political strategies and tactics are, in turn, linked to how the social materiality that constitutes the field of action for them is defined, its weaknesses and vulnerabilities.

From 2020, other actions will be proposed in the group: there will be the "30 actions for the climate" (the idea was to show many practices that can be adopted in everyday life, "good actions to reduce your own emissions", as defined by **S.B.**), events on sustainable tourism, training for adults—always with the aim of "involving people outside the network who already know certain topics".[8]

6.7 "I Cancelled Everything": A Pandemic Arrives

The health emergency made previous practices of confrontation and public exposure impractical. Swamped by personal problems, the ban on leaving the house and on demonstrating publicly, the activists began to assess the setbacks that had befallen the network. The members began to talk insistently about the need to restructure it. After a year of existence, the coordination of the groups loses momentum, participation decreases, and the euphoria of the last year gives way to the realisation that the difficulties are greater than the enthusiasm shown in the first year. According to **F.**, "the problem is that participation in the assemblies has waned"; it had to be revived by formally writing to the groups present to find out if they were interested in continuing to take part. But even this simple formality was not received without question. For **L.**, "there was no point in sending a communication without the existence of a project": "An e-mail saying that we are preparing a document for the municipality is a way of getting involved. To have a structural involvement, it makes sense to have a structural project". Collective articulation must be based on concrete proposals. The difficulties that emerge from concrete attempts

[7] https://www.lrb.co.uk/the-paper/v43/n22/james-butler/a-coal-mine-for-every-wildfire; https://newrepublic.com/article/162247/andreas-malm-blow-up-pipeline-climate-direct-action; https://www.thenation.com/article/environment/andreas-malm-interview/.

[8] Greenpeace milano, Manitese milano, ER milano, M4C, Resilient GAP, CTG, Diciassette, Climate Reality Project, CORE-lab, Climate Engage, Comitato Ciclamini Mimose, SeiGAS, IPR: "with the aim of involving as many people as possible in daily actions to reduce their emissions" (05/03/2020).

to organise collective actions that can combat climate change show us perfectly well that it's not enough to come together around a "common problem". For **N.**, "either there is something to be done together and people realise the importance of participating; but if the role is not clear, there is no reason to participate and they don't do it. We need to propose a project rather than just an email. I don't have a concrete proposal, but we need to think about it". According to him, the problem of participation, involvement and coordination of efforts needed to go beyond the mere formal existence of meeting spaces such as the one formed by the M4C. The existence of the M4C favoured dialogue, updates and collaboration (**F.** RG), but it was necessary "to understand who wants to dialogue with us, to understand what is happening in our 'choir'". **L.** pointed to an apparently participatory aspect of the network: "dialogue, it's true, but it's also true that those who have spoken have been few. Until now, they've never involved the network more than that. A structured project is needed".

After a year of intense demonstrations, activities and debates, and a few months after the emergence of covid-19, the "climate" in the group was one of doubt and uncertainty about its ability to carry out relevant actions. For **D.** (XR), "if the idea is to give life to this network again, this is a unique and crucial moment when the need for a united front for a common problem is most felt". For the activist, it was necessary to "catalyse everything on the basis of a common desire to raise the stakes, to create some kind of unity". According to him, this could be done, in the first place, through knowledge between the various individuals who made up the network. Therefore, "it was necessary to create a space for discussion and dialogue that revolved around a common theme: how to work for change?" For this reason, he proposes moments of dialogue that are not operational in nature, "but rather the circulation of ideas, something that can help them understand what they need to do and how to deal efficiently with the crisis". The proposal was received positively, but it didn't seem to be enough to take the network to the previous level; there was "the fear that the big associations (like Greenpeace and Legambiente) wouldn't want to take part" (**L.**), which could further diminish the network's relevance.

The activists found themselves in a difficult situation. The common problem did not offer "objective conditions" for a transversal, mass anti-politics. During the period following the first health crisis, we will see the multiplication of actions that pointed towards the deepening of territorial actions; they involved an aspect of alter-politics: mutual aid networks, such as the "Volunteer Brigades for the Emergency"[9] in which various student organisations (Tempo Zero, Resilient Gap) present in the M4C participate, and actions to create collective gardens.

The articulation between the groups, based on making their own differences explicit, became an essential movement to verify what possibilities existed for the continuity of the network: "But if we first try to strengthen ourselves by organising a meeting where we introduce ourselves and talk about the urgency, the problems and let off steam, it will be better. (…) Conclusion: we wrote emails with ideas from L. and P. with who we are and what we want to do, saying that we are invited to a next meeting to talk, discuss and renew ourselves and talk about the challenges we face now" (**F.**, RG).

How to keep these heterogeneous groups together? How to produce collective actions? These empirical accounts show us how far we are from assuming an "already given matter" on which it would only be necessary to imprint a particular form of political structure and organisation. The debates—and the difficulties—demonstrated by the Milanese activists take us to the heart of the political debates about what to do to avoid climate catastrophe and show us how the formation of a collective organisation that can tackle business as usual must be considered from another language, as an emergent property. A very active member of the M4C, **F.** (RG) concluded that "let's take the first step, which is to call the first meeting to understand who we are, to understand and bring out the desire to do (*far emergere la voglia di fare*)". Up to that point, according to **L.**, "within the M4C there were lots of things to do, but there was a lack of more substantial moments of confrontation". Protests, blockades, occupations and debate events took place intensively throughout 2019. The climate of mobilisation and engagement seemed closed. After a year of its creation, this was proving unrealistic in the face of resistance from business as usual and the health emergency. But this resistance didn't just

[9] https://www.facebook.com/brigatevolontarie.

occur in the "inactivism"[10] of politicians and companies but also manifested itself in the everyday lives of the population tied to fossil fuels for the reproduction of their livelihoods.

It was necessary to start all over again and, to do this, it was necessary to technically produce the means to articulate these differences to produce, once again, a force that could confront what seemed to be an ebb. At a meeting on 06/11/2020, the activists began a journey of questioning their own identity, "whether a subject or a platform", what they could and couldn't do and tried to establish "some rules of partnership, dissemination and participation" (**M** and **GP**). It was necessary to "create a narrative for the M4C" (**M.M.**) that would situate it as a reference and, at the same time, allow the various groups freedom of movement. The activists were unsure of what to do and how to organise themselves to continue, over the years, taking concrete steps in formulating responses to climate change and confronting the relevant public and private actors. The platform seemed appropriate for the moment and for its current strengths" (**C.** SaIZ); but if they managed to gain agility, at the same time the M4C "almost disappeared as an issue" (**M.**). For *A.*, "M4C adds value because it has the ability to act as a connector, without which that energy soon disappears". For **N.**, it was the opposite:

> "I think we always call ourselves a network, but we act very much as a subject (meeting once a week is an example). Before tonight I had more the idea of it becoming a subject but listening to you I changed my mind. Maybe the current phase is better that it becomes a platform. It means anyone can propose something, everyone proposes an idea, and the others decide if they want to participate or not. But we also need periodic (maybe monthly?) reflection moments where we try to think a bit more at a higher level (like municipal, etc.). I thought that, for example, next year there will be municipal elections and, for example, we could propose ten points to

[10] The term inactivism was coined by climatologist Michel Mann to define the new tactics of fossil fuel companies and politicians: "I call them 'inactivists'. They have stopped denying climate change altogether because it simply no longer has any credibility". We can all see that climate change is happening. Just look at what happened in Texas recently and other devastating impacts of unprecedented extreme weather events. This cannot be denied, so instead the inactivists have turned to a whole new set of tactics. Because, ultimately, they really don't care that we remain addicted to fossil fuels—as long as we remain addicted to them. Watch the full interview here: https://www.ews-schoenau.de/energiewende-magazin/zur-sache/michael-mann-i-call-them-inactivists/.

the candidates. I would combine a platform-like structure with moments of reflection."

The options seemed to fluctuate, it seemed difficult to find a "middle ground". **A.** (Ridef): "either we go towards a formal structure or a totally informal structure open to everyone". For **M.**, "if decided to become a platform, what would it do together? What would be the purpose of the meetings? If we are not willing to give up on the idea of a strong subject/network, even without putting the most active associations into action, they will be free to create something more independent".

Faced with the initiatives of more active groups, such as the FFF, the M4C activists note that the "network is no longer listened to" (**M.L.**). Low participation, according to **N.**, "is linked to the lack of concrete things, which is why people stop participating"; one of the possibilities was to create "moments of confrontation from which ideas are born and from which concrete projects can be born". The attempt to organise based on concrete objectives was seen as a viable solution for collective action:

"**M.**: As a goal, having a very generic objective (for example: what can we do to transform Milan in 10 years?) would allow us to connect much more with the reality of the city. This gives us much more capacity to understand and interact.
Q.: The target should be the city of Milan, because it is in Milan that strategic choices are made in terms of mobility, soil, energy and so on.
M.: Once you've set a goal, what matters is how you put it into practice. I'm not so convinced about setting a concrete and very specific goal, like the territorial committees. I opt for joint initiatives, not concrete projects, as has been proposed but not realised. The M4C is not very representative, we are not listened to, it makes more sense than reality, because the committees organise the initiatives and the M4C relaunches them. Now we don't have the strength to propose concrete projects.
N.: As we're a network of environmental associations, we can only do so much; by coming together, we can think of something that can have an impact on the citizen, from where ideas are born, and then those who want to participate, participate".

The movement/network/organisation is beginning to dissolve; initially, a way was sought to articulate and keep these heterogeneous groups together: it is no longer clear who can take part, everyone or just those who are part of a collective group. The meetings seem extremely formal, i.e. unsuitable for a group that is moving towards another format.

> "**M.L.**: What can we do at these meetings? It seems to me that there is someone who proposes and wants to take part, specific meetings can be held about this and communication channels can be activated. Meetings can be held about individual projects.
> **S.**: It's not enough to have a chat to see if the platform is working, meetings are for talking to each other, messages get through better, you can't network without meetings.
> **M. M.**: I prefer to call it a sharing space rather than meetings. I agree that there should be times when people on the platform can meet.
> **S.**: We could find a catchy name to relaunch the "meetings". If, for example, you only have one meeting a month, you could think about organising it with a certain quality.
> **M. L.**: What should we do in practice after the meeting?
> **V.**: define this objective, which is not a project (main objective), a meeting space that will no longer be called "meetings", but will be less frequent (21/06/2020).

But what could be done in this meeting space? According to **P.** (RG), they would give the opportunity to focus on "tactical objectives that concern the reality of the territory". It's a question of establishing an environment that allows actions to continue, while at the same time expanding beyond the existing groups.

> "**N.**: We already have our goal in M4C, we can start a meeting in September on, for example, "How can we start the ecological transition in Milan?", or, for example, "How can we make universities carbon neutral?", or, for example, in another meeting: "How can we push A2A[11] towards the transition?". Encourage everyone to reflect, for example, with one meeting a month.
> **S.**: I approve of what **N.** said, because it could help us to humbly put ourselves on the same level among the associations. Perhaps send extended meetings to the various associations.
> **M. L.**: define a list of channels in which the M4C will exist, a list of positions that must be filled (e.g. social channels), about the times of

[11] A2A, an Italian multi-service company, operates in the environment, energy, heating, networks and technology sectors.

the meetings, I like **N.**'s idea, he can be clever if we make these meetings coincide with the Klimatfest, on the days when the speakers will be speaking, and a week later the associations get together and draw up a sort of manifesto".

The question of organisation appears less as a question of applying a model (or a defined strategy) than as a question of establishing partial territorial *sets*: it is necessary to follow the various subjects, the problems they formulate (insofar as the situation questions them) and, through them, describe how they seek a transformation of existing conditions. For this reason, it seemed appropriate to follow the journeys of these activists to see how they sought to act collectively, but also how they sought to articulate specific practices that defined them as active subjects in certain areas. As the context remains open, we can observe how they try to delimit a space for intervention but what this whole journey shows is that there are no ready-made solutions that can be applied.

References

Dewey, John. (2016) The public and its problem: an essay in political economy. Ohio: Ohio University Press.

Hage, Ghassan. (2012) Critical anthropological thought and the radical political imaginary today. *Critique of Anthropology*; 32(3): 285-308.

Knox, Hannah. (2020) *Thinking like a climate*: governing a city in times of environmental change. Durham and London: Duke University Press.

Latour, Bruno. (2018) *Down to earth*. Politics in the new climate regime. Cambridge: Polity Press.

Malm, Andreas. (2021) How to blow a pipeline: learning to fight in a world on fire. London: Verso.

Simpson, Michael. (2020) Fossil urbanism: fossil fuel flows, settler colonial circulations, and the production of carbon cities, *Urban Geography*, 01–22.

Stengers, Isabelle. (2015). *Cosmopolitiche*. Roma: Luca Sossella editore.

Stengers, Isabelle. (2020a) *Réactiver le sens commun*: lecture de Whitehead en temps de débâcle. Paris: Les empêcheurs de penser en rond.

Stengers, Isabelle. (2020b) We are divided. *e-flux journal*, issue 114, December, 01–06.

Swyngedouw, Eric. (2010) Apocalypse forever? Post political populism and the spectre of climate change. *Theory, culture and Society*; vol.27(2–3): 213–232.

Viveiros de Castro, Eduardo. (2019) On Models and Examples Engineers and Bricoleurs in the Anthropocene. *Current Anthropology*; 60:S20: S296-S308

CHAPTER 7

Destabilisation and Disruption Practices

> "A world revolution for climate justice has no clear historical precedent (...) we have no previous model to go by" (Mann and Wainwright 2018: 178)
>
> "Tout se passe comme si la densité des modes d'existence, la pluralité des relations que nous pouvons entretenir avec les êtres qui forment nos milieux et qu'ils entretiennent entre eux, l'hétérogénéité de nos savoirs, cette « surabondance du réel » étaient la cible d'un leitmotiv: une seule logique pour l'hétérogénéité des savoirs, un seul mode d'existence pour la pluralité des êtres, un seul cosmos pour la diversité des mondes." (Didier and Stengers 2021, p.130)

When analysing the experiences of becoming active, we must stick to what "communicates with the immediate facts"; however, it is not a question of claiming these facts as having a privilege over others; it is simply a question of observing that there is experience and that understanding is therefore necessary (Stengers 2015: 711). A simple search in one of Italy's

D. Delatin Rodrigues, *Climate Change and Political Action in Italy*,
https://doi.org/10.1007/978-3-031-98771-7_7

most widely circulated newspapers, with local sections on Milan, demonstrated this. The year 2019 alone has the same number of mentions as the years 2005 to 2017 combined.[1]

Events multiplied with such intensity in 2019 that it was difficult to keep track of them all. "The year 2019 was the momentum" (S., personal communication, June 2020)—a momentum that would be drastically interrupted by the outbreak of the Covid-19 health crisis in February 2020. The issue of climate change began to be discussed in numerous spaces, always from a critical re-reading of the usual practices, anchored in the available climate data. Participants identified themselves as climate activists and, from this position, sought to create new attachments—material, affective, political—that would allow them to "break out" of fossilised socialities.

This process was not restricted to the younger generations, as is commonly assumed. The "climate" of open debate favoured intergenerational encounters and clashes of perspectives. In December 2019, for example, I took part in an event where members of the Gruppo di Acquisto Solidale (GAS) in Lambrate invited local members of Extinction Rebellion to discuss ideas, practices and strategies for tackling the climate and ecological crisis. The older members, who called themselves "elders", expressed scepticism about the effectiveness of isolated civil disobedience actions. For them, it would be necessary to build a mass political movement.

The most revealing point, however, came in the form of a question: one of the GAS members asked the young people from XR how they bought their food, whether they took part in any solidarity consumption networks, and what kind of markets they went to. Faced with negative answers—based, at the time, on dependence on large-scale distribution systems—the GAS member pointed out that this was a central aspect to be taken seriously: "How can we dismantle business-as-usual if we continue to feed it with uncritical consumption practices?".

[1] Quick search with the key words: climate change—Milan. Period: 2021(112) 2020(159) 2019(180) 2018(59) 2017(54) 2016(26) 2015(37) 2014(23) 2013(16) 2012(14) 2011(10) 2010(14) 2009(21) 2008(22) 2007(11) 2006(6) 2005(2).

7.1 Unlocking Fossil Sociality

But what was the activists' goal anyway? It would be difficult to synthesise it into a single formula—which is precisely why it is more productive to observe the points of collision where disagreements over methods of action become visible. As we shall see, the shared diagnosis of climate change as a problem is not enough to unify perspectives on what to do. In problematic situations, we must pay attention to critical moments when doubts emerge about the meaning of action: how to act coherently in a world marked by the climate crisis, where every practice can be interpreted as contributing to its acceleration or deceleration?

It is in this context that we should consider the existence of an "ecosystem of practices and tactics". Many of the practices followed by climate activists in Milan are not unique to the city; they are part of a wider repertoire that has been developed and adapted locally. Called "practices of disruption" (Grasso and Delatin Rodrigues 2022) elsewhere, these actions are carried out in what Debaise (2017: 56) calls a "local unit of experience". The subject who engages in them is also produced in this process: their subjectivity is constituted step by step, entangled in fossil devices, extractive territories and chains of consumer goods. Although it cannot be reduced to these structures alone, its emergence is linked to them.

There is therefore no single strategic direction for breaking with fossil domination. There are multiple points—material and subjective—where the fissure can begin. In this section, I have selected some cases that show how Milanese activists are trying to confront the almost uncontested hegemony of Italian fossil fuel giant ENI (Ente Nazionale Idrocarburi) in the Italian public debate on the climate crisis and the energy transition.

This confrontation, however, is complex: it is not reduced to denunciation or frontal attack. ENI's presence cuts across multiple dimensions of social life—and its strength derives not only from the continuity of its operations in the market, but also from its ability to build legitimacy. In other words, to reproduce consensus around its "beneficial" role. It is precisely at this point that the action of activists becomes crucial.

Against rampant consumerism, they organise clothes swapping events (Block Fridays—Make Friday's Green Again) and local food markets (Gruppi di Acquisto Solidale, like the one in Lambrate). In reaction to the expansion of cemented urbanisation, they get involved in the production of agroforestry (Parco della Vettabbia) and the defence of threatened green areas (Parco Bassini, Goccia, Baiamonte). In response to the inertia of urban mobility policy, they block traffic with bicycles and their own bodies (Bike Strike). They occupy the headquarters of RAI (Radiotelevisione Italiana) to denounce the insufficient coverage of the ecological crisis. They protest in front of ENEL and demand energy transition plans. They intervene in the annual meetings of ENI and ENEL in Milan with direct actions and demonstrations.

In this chapter, I will present these episodes in the form of vignettes—brief descriptions that illustrate how breaking the fossil assemblage involves subjects in decisions that directly interfere with the unequal conditions of personal and collective reproduction. These actions produce emotional, ethical and political trajectories that cannot be understood simply as local translations of global knowledge. They articulate diverse groups, institutions and urban and extra-local spaces, in a time span that goes from 2019 to 2021. The information was gathered from mixed sources—social media (Facebook, Twitter), messaging apps (WhatsApp, Telegram) and participant observation since the beginning of 2019.

7.2 Cross-Cutting Policy?

Two months after the largest climate demonstration in the city of Milan (15/03/2019), the activists gathered in the *Milano per il Clima* network discussed the pertinence of taking part in an initiative that sought to contrast the rise of nationalism and racism in Italian public life. The event was called *Gran Galà del Futuro*, organised by Nemumadimeno[2] and *Milano antifascista antirazzista meticcia e solidale*.[3]

The event was supposed to coincide with the closing of Matteo Salvini's election campaign on 18 May 2019 and would be attended by the well-known French far-right leader Marine le Pen in Milan. According to the organisers' statement,

> "All those who cultivate the dream of a return to fascism, strong national borders, the supremacy of the 'white race', the condemnation of homosexuality and transsexuality and the return of women to the home to look

[2] https://nonunadimeno.wordpress.com/2019/05/17/18-maggio-gran-gala-del-futuro-indietro-non-si-torna/.

[3] https://www.facebook.com/milanometiccia/.

> after the family or to brothels, reinforcing the image of the mother or the prostitute, will be present.
> While an entire generation is on the streets denouncing that climate change is taking away their future, European right-wingers promote the unconditional exploitation of land, bodies and territories, foment war between the poor to avoid class conflicts, promote the persecution of gypsies and Sinti, deny universal rights by failing to save lives at sea, fund countries like Turkey and Libya that exploit and torture migrants and open administrative detention centres.
> To fight this reactionary wave, we want to organise a day on 18 May that will take fabulous Milan to the streets to say no to the return of nationalism and all forms of fascism in Europe and the world: we want to bring the GRAN GALA DEL FUTURO to life, because there's no going back!".

Despite the apparent support of most of the groups gathered there, some objections were raised. Firstly, the method was questioned; insofar as support was not unanimous, joining the campaign could not be done in the name of the M4C network; according to the account of one of the participants (a member of a student collective at the University of Milan (Statale)

> "First of all, we are a network and joining as a network implies that all the member organisations (or at least the majority) have expressed their support. That hasn't happened, as we well know; and as we haven't yet set up the working groups, there hasn't even been a decision by an internal M4C body that is delegated by everyone to take such a decision. So if those who are taking part in the current discussion make that decision, the others won't feel represented and won't really feel part of Milan for the Climate".

Secondly, the campaign's commitment to climate and ecological issues was questioned: for some activists, it seemed that the campaign sought to instrumentalise the climate and ecological crisis to attack the ruling group in the Lombardy region (the right-wing Lega in the first place), with which some of the members of the M4C network had institutional relations. According to a message posted on the network's messaging chat

> "However, beyond the question of participation, M4C is a political network, but not a partisan one. This means that we take political positions on our issues, the environmental ones; but the event on the 18th seems to me to be more against specific political formations than in support of a

specific issue. It's true that in the description there is a reference to climate change and it's also true that some (not all) sovereign realities are denialist; but, frankly, it seems to me that the thematic centre of gravity of the event is not the environment. Therefore, taking part in the event implies, in my opinion, unconditionally taking sides against the League and its allies (who also govern in the Region and with whom we have an open table) and this, in my opinion, goes against the non-partisan nature. A demonstration on environmental issues against the government would be different, because in that case there is a link with our activism, but it seems to me that this event doesn't have many environmental issues".

Other associations reinforced this assessment by explicitly requesting that the initiative not be publicised through M4C's official communication channels. The recurring reason was the political nature of the event: "I wouldn't like the political aspect as an association" (E., Naturiamo). The division between what would be a strictly climatic sphere and other political spheres was the main point of disagreement—a disagreement that would reappear at various times during the network's collective activities.

For some, this separation was strange, as they considered that issues such as climate change, economic exploitation and migration were inextricably linked, and that this intersection would only tend to deepen. Even so, the distinction between the "climatic" and the "political", held by certain groups, helps to outline the impasses faced by Milanese climate activism: how to build a common perspective on what to do and how to act in the face of the crisis?

By distinguishing climate issues from explicitly political agendas, certain groups reaffirmed the *motto* of "following the science". For them, this stance would allow them to move directly from the data provided by climate models and scientific projections to government action—thus maintaining a position of technical neutrality and objectivity.

While some student organisations—especially those closer to left-wing social centres—defended the importance of officially participating in the anti-racist demonstration, groups with more traditional structures, such as NGOs and environmental associations, preferred to keep their distance, reinforcing their identity and strategic separation.

From this divergence emerged a broader confrontation over ways of conducting climate politics "from the bottom up". On the one hand, there were those who believed in the effectiveness of pressurising the current political system, operating through existing institutional tools. On the other, groups that advocated the creation of parallel, autonomous spaces in which to experiment with alternative political practices. A clear manifestation of this difference occurred during the Covid-19 pandemic, when several "solidarity brigades" were formed by activists, including climate activists, to help vulnerable populations—an action that escaped traditional institutional moulds.

This is not about resorting to simplistic dichotomies. Rather, it's about recognising two distinct—and coexisting—diagnoses in the field. One believes that the climate response can be driven by reforms within existing structures, if there is sufficient social and political pressure. The other considers that the current immobility is structural, rooted in the complicity between the state and the fossil industry, and that it will therefore only be tackled by building new forms of deliberation and participation.

Between 2019 and 2021, various experiments were carried out in the city to demonstrate the viability of "bottom-up" solutions: the installation of solar panels in low-income housing by activists linked to social centres, sustainable self-construction laboratories, community gardens and self-production food systems. The point here is not to judge their ability to reverse the crisis globally, but to recognise the practical effort to reconfigure everyday habits based on intensive energy consumption and dependence on extractive chains, both local and distant. For some Italian collectives, these practices represent paths to "exit fossil".

By refusing to take part in the anti-racist demonstration, certain groups sought to keep the climate discussion restricted to environmental and energy aspects, guaranteeing their presentation as neutral, technical and non-partisan interlocutors. This difference was also clear in the way the big public demonstrations were organised. For some, they had to be carefully choreographed to generate communicative performances consistent with the cause's institutional message. Posters, banners and slogans should allow "people of all ages to publicly express their opinions" (D., 35). For others—especially those linked to more radicalised collectives—this seemed necessary, but clearly insufficient.

7.3 Carrying Out Mass Demonstrations

In the run-up to the next demonstration on 24 May 2019, some information began to circulate about the event. The idea was to launch some balloons during the demonstration so that messages could be transmitted in them, which could hardly be done simply by climbing the buildings.

The problem, said one of the organisers, "was how to get to a height of 8 metres without risking death or having to climb stairs". Having been approved at the meeting on 14 May, the other participants were told that the balloons had been ordered. This announcement triggered a long exchange of accusations about the incoherence of the choice. The first to speak out was a member of an important ONG:

> "Excuse me, but when were the balloons decided? That really is a contradiction to our fight. We're all fighting against single-use plastic, and we turn up with 200 plastic balloons? Sorry, but no. Are we going to stay consistent?"

How can the message be conveyed without using means that could contradict the message itself? How do you accept a decision made by the majority present at a meeting that seems to contradict minimal concerns about the impact on the environment and people's assessment[4]?

> "We were told that it would be biodegradable plastic and that we would do our best not to disperse it. I was sceptical too, but that's what the majority decided on Tuesday. I was sceptical too, but that's what the majority decided on Tuesday. I was sceptical too, but that's what the majority decided on Tuesday".
> "I can tell you that it's normal to disagree, but it's also normal for decisions to be made at the meeting, with those who are there".

The representative of another important Italian ONG says: "After a quick collection of opinions, we says no to balloons. It's precisely the disposable concept that's wrong".

Those who proposed the action justify it by saying that they are biodegradable bladders and will be collected at the end, but nothing that can placate the growing condemnation of the choice:

> "I'd just like to emphasise that biodegradable doesn't mean not harmful: an animal that eats one of these balloons dies without too much trouble, and we expose ourselves to a huge media risk by putting balloons out there..."

[4] "And above all, once the action has been carried out, both the balloons and the posters must be removed so as not to risk criticism." (**E.**).

It was necessary to maintain "coherence" between the message, the media and the relationship with the public,

> **E.**: Balloons Elio, all things highly polluting, I know you're doing the best you can, but a message of incoherence can get through.
> **G. P.**: We have to think of some kind of action. The posters don't seem to have much impact.

These small gestures make explicit the differences in which public action must be taken. An attention in which the message about the climate catastrophe must be carried forward without the use of fossil materials. The same dynamic occurred with an artistic performance at the 19/09/2019 demonstration in front of Milan's Duomo, where a representation of the globe was set on fire. The action provoked the fury of some and further demonstrated the network's internal differences.

One of the main points of tension within the Milano per il Clima network concerned the position of Fridays for Future (FFF). As an autonomous organisation and the undisputed organiser of the morning demonstrations, FFF occupied a central position that often forced M4C to adapt its own strategies: even though it sought to chart a more inclusive path, the network ended up orbiting around FFF's decisions or reacting to them.

As J. (Tempozero and member of the FFF) explained, the logic was simple: "The FFF has open assemblies and if these proposals are approved, they are carried out". These seemingly minor moments make visible the uneven distribution between what is treated as a simple "organisational fact" and what constitutes a real "political issue". They make explicit the problematic materiality of politics: not only does politics deal with material issues, but it is itself constituted and mediated by objects, performances, infrastructures and organisational formats that are not neutral or given.

These small vignettes help to explore the variety of dimensions that make up climate activism, between anti-political and alter-political practices. However, it would not be possible to establish a rigid taxonomy between the groups based on their responses to problematic situations. The positions oscillate. As the following vignettes will show, the main internal divide revolved around the acceptance—or not—of direct action, and the definition of what counts as "violent action".

Let's take the use of objects in public performances as an example: while some groups defended visually striking actions, such as the use of balloons or the staging of a globe on fire, others warned of the communicational and symbolic risks of these choices. For the former, the strength of the message justified small material impacts or possible media discomfort. For the latter—usually NGOs or associations concerned with the coherence of their public image—these means jeopardised the effectiveness of the cause by generating noise and contradictions. "We have to maintain coherence", they said, fearing that the dissonance between means and ends would undermine the movement's credibility.

The criticism was clear: presenting scientific data is not enough. Consensus on the need for climate action needs to be built—but how? Two strategies emerged. The first relies on clear, coherent and scalable communication, on the idea that by reaching a sufficiently large number of people, there will be a "tipping point" capable of transforming climate policy. The second emphasises disruptive, direct action, even if this implies costs in terms of public support.

This divergence ultimately reveals two distinct logics of political practice: one gradualist, which preserves the functional divisions between spheres (institutions, civil society, media); the other disruptive, which seeks to cross these spheres, forcing their reconfiguration through external actions that destabilise normality.

However, this is not specific to the situation in Milan. The dilemma over how to act politically in the face of the climate crisis cuts across national, regional and local contexts. As many analysts have pointed out, we live in a paradoxical situation: the scientific evidence on the severity of climate change is increasingly robust, but the political subjects capable of confronting and reversing the hegemony of the fossil complex remain absent or weakened (Latour 2017; Mann and Wainwright 2018; Danowski and Viveiros de Castro 2014).

As business-as-usual continues its course, accompanied by increasingly alarming warnings, urgent questions arise: How to break with the material and symbolic hegemony of the fossil bloc (Grasso 2022)? How to expand engagement in the climate movement? What is legitimate to do, and with what means?

7.4 To Occupy or not to Occupy?

But how can we put pressure on the operational agents of fossil sociality? Those responsible not only for extracting fossil fuels, but also those who distribute, sell and make them consumable? The apparently limited scope of these activists' actions, however, touches on important dimensions of the political debate about what kind of action should be taken and, perhaps more importantly, what actions are considered legitimate. What is the limit of political action? Legality, legitimacy? Dialogue, confrontation? On 22/05/2019, FFF activists, including members of the student collective Tempo Zero—one of the organisations that joined M4C—announced.

> "We're occupying Enel's headquarters on Carducci Street, on the corner with Cadorna, and anyone who can should come and support us, even if it's just to say hello. We'll probably sleep in there and spread the word".
> "Things to take to Enel as soon as possible: banners, tarpaulins, used posters, spray cans, PowerBank, food, water, today's printed communiqué and 24 leaflets, sleeping bags, whisky and string."

Immediately, several groups began to take a stand against the action, many of whom called for the action not to be published on M4C's communication channels, and others particularly disassociated themselves from the method, the choice of target and the date chosen for the action.

> "Guys, Mani Tese is withdrawing its sponsorship of this event. We don't agree with Enel's occupation".
> "The Climate Reality Project has also disassociated itself from the event."
> "Msoi is withdrawing its sponsorship of the event. The action was not discussed at the meeting."
> "Also for Resilient GAP, the action should not be sponsored by the page"

The rejection of the action surprised the promoter, who tried to justify himself by saying that it was an autonomous action and that there was no association with M4C.

> "It seems to me that we're not worrying about anything. I understand the objections of those who are against it (I also have some, although I think the action is justified), but I don't feel that anyone is associating M4C with the occupation of ENEL. And anyway, even dissociations, as long as

they remain in that group, can serve to reaffirm a position, but they don't reach the public".

The fear of some M4C members was that, with an action of this kind, they would effectively scratch themselves, something like a "noexpo due"—the 2015 demonstrations against the universal exhibition held in Milan that year, in which clashes on the roads generated public protests what was classed as vandalism—rather than antagonism. According to several members, the fear was to see an action considered violent associated with the whole climate movement in Milan. The way this gesture would be narrated by the media and received by the general public was a permanent cause for concern.

> "**D.** (Climate Reality): Without wanting to perpetuate eternity, and seeing (in my personal opinion) that Enel's choice is questionable, I wonder what one can agree with, but anyway. There is one detail. The headline says "occupation by emissions and due to strike". So there is an association with the strike, yes, but anyway..."

On 26 May 2019, the European Parliament was symbolically occupied by an international group of FFF activists and the activists responsible for M4C communication (members of Tempo Zero and FFF-MI) published the news on the group's Facebook page. The dynamic is repeated: **D.** shows his disagreement and accuses the Tempo Zero members of monopolising communication for specific political ends—that of giving visibility to the FFF and promoting tactics not shared by the other members of the group:

> "He is using his position in a propagandist way by not guaranteeing equal access to social services"; the practice of occupation was unacceptable, especially in an institution "that cost the blood of two wars"
> **F.C.**: "But where did these people go? I mean, from the photos you can see, it looks like they've set up a tent, like in front of the entrance. But what are they doing?".
> **D.**: "Occupation is a method that is not fully shared by everyone."

When asked why the occupation could not be considered a valid tactic of political struggle, **D.** argues that, firstly, it is illegal and, secondly, it is an end in itself. But not everyone saw it as a violent attempt to impose their own will against the institutions:

> **M.**: "Well, I'm going to give my opinion on the news that has generated all this controversy. As far as I can see, the FFF has not occupied the European Parliament, preventing work (which is not taking place at the moment), but only the space in front of it. So it seems to me that they are behaving like a movement that is acting to draw attention to a specific issue and that seems totally legitimate to me".

Discussions about how to organise the next demonstration are confused about how to reconcile what has become obvious to everyone. At the same time as the Venice Film Festival was taking place in 2019, a group of climate activists decided to take advantage of the situation of high international public visibility to carry out actions calling for climate change to be challenged immediately: a member of the group wrote a long report that we consider appropriate to leave in full:

> **G.** (M5S): Yesterday, at the Lido, I got to the courtyard of the film screening and found the demonstration. They rhymed the slogan "we are all anti-fascists", then criticised the new government and insulted the police. There was a banner in support of immigrants and dozens of banners from the "no big ships" committee. (...) Banners about the climate? I saw very few. I wonder what idea the people who were there had of this "event", especially the people from the Veneto region, who voted for the denialist Salvini. In any case, as the risk of being exploited and slandered by the press is well known, it seems to me that the FFF and the climate movement in general come out rather damaged from this rather chaotic mix with other demands, some of which are worthy in themselves, and especially from this homologation with a certain left and its methods. The deniers have a great day: environmentalists are the same as communists. For these reasons, yesterday's demonstration, which partially blocked access for a few hours and caused some discomfort, didn't strike me as a good gesture towards an institution that is trying to contribute, in its own sphere, to a better world. If the aim of the demonstration was to gain visibility in the media, exploiting an event of international importance in a somewhat opportunistic way, it seems to me that this should never be the attitude and style of the climate movement, which, more than ever, needs to arouse sympathy and not mistrust around it. It was billed as a climate march and it should stay that way. In retrospect, if during the eleven days of the exhibition some gazebos had been set up in the area adjacent to the screening rooms, with well-chosen information material, it would have been possible to sensitise many of the people from all over the world who stop in the

area with nothing to do in the intervals between one film and another" (chat - 09/2019).

It is precisely a total refusal to consider as valid any act that could be considered disruptive by those who suffer it: blocking traffic, insofar as it prevents motorists from passing, would be unacceptable because it impedes and creates difficulties and discomfort for those who need to get to work, home or simply move freely around the city; this intervention is followed by other comments about how change can only take place peacefully.

> "Non-violence (as opposed to physical psychological verbal violence against people or things), democracy and freedom of expression. If that's missing, I don't think there are any extenuating circumstances or more or less different interpretations or points of view".
> "The problem is that these are conceptually questionable forms of struggle. The great lesson that history has taught us is that non-violent struggle leads to much more significant results than violent struggle. If we want to have a future, we can't ignore this factor. Most people, fortunately, still condemn violence in all its forms" (chat - 09/2019).

But not everyone was unshakeable. This was "the hallmark of M4C" and probably the reason for its break-up:

> "I think that's the wrong approach to the problem. If 'people' make these questionable equations, you need to work to educate them, not to attack those who follow antagonistic practices that, even if we don't agree, have the same goal as us. If you like, it's the same, in reverse, with dialogue with "green" parties/institutions/companies: you may find this path futile and/or even harmful, but it makes no sense to attack those who do it, regardless of who is right, and who are fighting for the same reason. Arguing about who is better between x and y is old and sterile, and only plays into the hands of those who are against both" (chat- 09/2019).

7.5 Fossil Culture

ENI seems unattainable. With the exception of a few critical reports, it is rare to find articles in the mainstream press that openly question the contradictions between the energy transition discourse and corporate practices based on business as usual. ENI's innovation sector, in fact, has a regular column in the sustainability section of Corriere della Sera, one of Italy's largest newspapers.

Faced with this narrative hegemony, activists from different backgrounds have started to demand more attention from the media—especially public broadcasting—when it comes to covering the causes and consequences of the climate crisis. On 20 September 2019, members of social centres (such as Lambretta), student collectives (Tempozero), Fridays for Future—Milano (FFF-MI) and Extinction Rebellion—Milano (XR-MI) symbolically occupied RAI's headquarters in Milan. The slogan of the protest was "Tell the truth—Act now—Change the system" (Dire la verità—Agire subito—Cambiare il sistema). In a statement signed jointly by FFF and XR, the activists declared:

> "Public television has begun to deal with the problem, but it has a duty to tell the truth, abandoning the emergency and contingent rhetoric that dominates the current narrative. All news, at this historic stage, needs to be contextualised by the climate crisis. Everyone who accesses the public service has the right to know what is at stake - and what can be done."

However, the action generated new divisions within the Milano per il Clima network. For some members, the RAI occupation was perceived as excessive and counterproductive. One activist summarised the feeling of disapproval forcefully:

> G. (M5S): "The same thing happened with the protest at the Venice Film Festival. If you insist on holding one protest a month, at least choose the right targets! There are many other oil companies besides ENI... and car manufacturers who are selling electric models at absurd prices. Please, more clarity!".

Faced with this criticism, another member of the network responded:

> "No one today denies the enormous power that the media wields. If we agree that the climate crisis is urgent and terrible, what's wrong with demanding that it be treated with the necessary seriousness? If they prefer to act from within, seeking meetings and negotiating tables with the RAI, let them do so - as individuals or as associations. But let's not confuse different fronts of action or oppose them. To do so is to play into the hands of those who oppose them all."

A year later, on 21 October 2020, the Milanese chapter of XR returned to the scene with another symbolic protest. During an event promoted by the Fondazione Giangiacomo Feltrinelli—which featured philosopher Timothy Morton and discussed the climate and pandemic crisis—activists interrupted the programme to denounce ENI's sponsorship of the foundation. They read aloud a statement demanding that Feltrinelli break its ties with the oil company, arguing that such a partnership gave ENI undue visibility and legitimacy. The appeal was clear: "use your prestige to put pressure on ENI".

The foundation's response, however, was firm. It claimed to be a "free cultural institution, open to all points of view—including those in favour and against the use of fossil fuels". It emphasised that ENI's resources have been used to "produce studies, hold events of cultural relevance and pay various researchers", and that this has helped to consolidate Feltrinelli as a reference in the national and international public debate. The foundation concluded by reaffirming its mission and rejecting any attempt at interference.

This is a point worth emphasising: the same activists who organise alternative food distribution markets are also those who promote street demonstrations and direct actions against the big oil sector. There is no split between the micro-local and the macro-structural, between the personal and the systemic; the crucial question is how these levels are articulated. To the well-established slogan "change the system, not the climate", we can add—as a gesture of problematisation—the statement of an activist during the first national assembly of Fridays for Future, held in Milan in April 2019: "we are the system". In this specific situation, the phrase should not be interpreted as adherence to the generalising notion of the Anthropocene—which distributes responsibility equally among everyone—but rather as recognition that "we" are interwoven in the material and subjective chains that reproduce fossil socialities.

Breaking with the omnipresence of fossil fuels therefore includes the body and its supporting materialities; interrupting their circulation is thus a central dimension of political action. The vignettes presented throughout this work reveal how activists transform specific situations—ranging from domestic spaces to public institutions—into occasions to highlight their everyday complicity with business-as-usual. Becoming active in a problematic situation is, in this sense, the result of a continuous process of clarifying the conditions that sustain their habitability in the first place.

References

Debaise, Didier. (2017) *Nature as event*: the lure of the possible. Durham and London: Duke University Press.

Debaise, Didier; Stengers, Isabelle. (2021) Résister à l'amincissement du monde. *Multitudes*; 85: 129-137.

Grasso, Marco. (2022) *From big oil to big green*: holding the oil industry to account for the climate crisis. Cambridge: MIT Press.

Grasso, Marco, and Daniel Delatin Rodrigues. (2022) "Disrupting to Decarbonise Socio-Energy Systems: The 'Carbon Transformation Axes' Framework." *Energy Research & Social Science* 90: 102657. https://doi.org/10.1016/j.erss.2022.102657

Latour, Bruno. (2017) *Facing Gaia*: eight lectures on the new climate regime. Cambridge: Polity Press.

Mann, Geoff and Wainwright, Koel. (2018) *Climate Leviathan*: a political theory of our planet future. London: Verso.

Stengers, Isabelle. (2015b). *Cosmopolitiche*. Roma: Luca Sossella editore.

Viveiros de Castro, Eduardo; Danowski, Déborah. (2014) *Há mundo por vir?* Ensaios sobre os medos e os fins. Florianópolis: Cultura e Barbárie/ISA.

CHAPTER 8

Become Active at the End of the World: Regeneration and Care

> "The present of extra-modern peoples may be a prefiguration of everyone's future." (Viveiros de Castro 2019, s304)
>
> "Régénérer n'est jamais une affaire générale, car il s'agit de créer ou de réactiver des rapports de proche en proche, toujours tentaculaires, toujours partiaux, toujours à cultiver, c'est dire à reprendre sous le signe de l'absence de garantie, et aussi de la douleur lorsque le perte est irremediable". (Stengers 2020:196)

When I talk about imaginaries of the end of the world, I'm not only referring to the multiplicity of fables about this end, but also to the multiplicity of worlds that could be ended. In order for these imaginaries to become explicit, they need to be situated—and the reverse is also true: in order to be situated, they need to be understood in terms of the whole process that creates them—and attached then—and it is on the basis of these imaginaries that attempts are made to establish new conditions for personal and collective life. They jump without knowing if they'll find something to hold onto.

The end of the world is narrated based on premonitory evidence—events from the past, the present and, paradoxically, the future, such as those that appear in predictive climate models. These imaginaries also operate as markers of belonging. In the following case study, I investigate two collectives organised around practices that the participants themselves

D. Delatin Rodrigues, *Climate Change and Political Action in Italy*, https://doi.org/10.1007/978-3-031-98771-7_8

define as socio-ecological regeneration. These are practices that attempt to formulate the **climate crisis as a problem of inventing new material bonds (Latour 2018).

I'm interested here in the imaginary of the end of the world as a socio-technical imaginary—that is, one that is not limited to moral statements about how the world is or should be, but is rooted in the materialities and technologies that mould the concrete horizon of action.

Instead of a temporality of the end, what is at stake is a spatiality of the end—the end of the world understood as the exhaustion of relational spaces. It is this spatiality that becomes the target of political practices, understood here as efforts to reconfigure territories in the face of crisis and extinction. The spatial apocalypse is revealed in the corrosion of bonds and ecosystems, desertification, salinisation and the production of what Saskia Sassen (2010) calls "dead land"—territories purged of any relational or reproductive value.

In this way, the daily reproduction of high-carbon socialities—the ways of life and spatial arrangements produced during the "Great Acceleration"—appears as a continuous process of impoverishment of living spaces for humans and non-humans. As XR activist Alice puts it, it's a "toxic" environment. It is precisely from this toxicity that the activated subjects seek to free themselves. They recognise that this toxicity has led to "people's dependence on fossil fuels", and that a process of detoxification is now necessary—"a rehabilitation" that allows them to relearn how to live without the continuous use of these fuels and their derivatives.

I define this group of professionals as climate activists. However, instead of dwelling on the political-ideological strata that make up Milanese activism, I chose to analyse the effects generated by a specific practice that explicitly presents itself as a response to the climate crisis: the regeneration of socio-ecological relations through the creation of gardens and agro-forests in two areas on the southern outskirts of Milan.

The grammar of regeneration seeks to function as a relational analytical tool that inscribes subjects within the broader economy of fossil socialities. This is not, of course, about defending regeneration as the definitive solution to climate and ecological problems. Rather, it's about taking seriously what those who become active in concrete situations say and do—how they interpret climate and ecological events and how they project forms of response, care and anticipation from them. As Stengers (2011: 281) suggests, "Each cry demand the recognition that it imports, and there is a speculative obligation to give meaning to that demand".

The actions of these subjects can be situated in what Schlosberg and Coles (2015) call the new environmentalism of everyday life: a form of engagement that articulates material and immaterial, subjective and infrastructural dimensions, and which manifests itself as material participation (Marres 2015). Their practices show that the materiality of social life—objects, technologies, non-human living beings, energy flows—can be understood as an assemblage (Deleuze and Guattari 1987), that is, as arrangements that could induce specific forms of relationship and subjective emergence.

This doesn't mean that these activists lack a long-term strategic vision. Perhaps it would be more appropriate to say that their perspectives on political transformation are reconfigured in the light of the horizon of extinction and ecological collapse—which leads them to invest in creating spaces of care and regeneration. By defining themselves as agents of regeneration, their activation is codified in healing practices that are not limited to humans but also include attached with more-than-human agents.

In this context, the agroforest or garden becomes the main aggregation device for regenerative collectives, acting as a nucleus around which other practices are articulated: self-construction, sociability, mutual learning. These individuals are immersed in forms of action that seek to increase both their effectiveness and the more-than-human agents. The main effect of the garden or agroforestry, therefore, is to function as a device that intensifies resonances—ecological, affective, social—and multiplies the possibilities of composition. In this sense, it is a generative device, in the terms of Debaise and Stengers (2021: 136): "modes d'agencement intentionnels, fabriqués collectivement, qui tout à la fois présupposent et induisent la capacité de celles et ceux qui y participent de faire sens en commun à propos de situations qui les impliquent".

The two experiences[1] presented here are not equivalent, as they are different cases even from a legal point of view—one taking place in an

[1] These are two groups that I could follow from April 2019 to December 2021 with different intensity and presence. With the Regeneration Heroes group, I was able to carry out lengthy participant observation, effectively involving myself in the group's activities and sharing different moments of interaction and dialogue. The second was through a few visits to the area and material published on the Internet. My information came mainly from a participant I met before the start of my research activities.

area granted by the municipal administration and the other through the occupation of an abandoned area owned by *Ferrovie dello Stato*.

But there are other elements that help us understand more precisely what is at stake in the practices observed in the two cases. Rather than operating from a pre-defined plan with the aim of modifying the "ecological mesh" (Kohn 2021), these collectives engaged directly with it, seeking to complexify it from within. It was less about imposing an external design and more about immersing themselves in the already existing interweaving's—human and non-human—in order to activate their transformative power. This didn't mean simply introducing new species, but reconfiguring spaces of more-than-human sociality through assemblages with local materials, collective practices and affective bonds. The aim was not only to exemplify the possibility of living outside the chains of fossil dependency, but also to establish, in the making, forms of cohabitation guided by principles of interdependence and reciprocity.

I follow these practices as situated gestures, in which both the re-composition of more-than-human agentivity and the opening up of human subjectivities to new experiences of becoming-with are at stake (Haraway 2016). It is at this point that we can draw a distinction between the "climate citizenship" promoted by the resilient city—functional, predictable, orientated towards adaptation—and the subjectivities that emerge in these collectives: ways of living that are open to the unpredictability of local and extra-local interdependencies, which refuse to be reduced to the logic of risk management and efficiency.

In this experimental space, other practices find an environment where they can emerge and be welcomed—information about wild plants that can be consumed[2]; cultivation techniques; construction, mobility; knowledge is shared and new skills are acquired. Instead of the notion of awareness, it might be better to use the notion of contamination and imitation, insofar as the person is affected and needs to learn, physically and affectively, to situate themselves in an environment full of intentionality and consequences.

[2] https://fresche-ribelli.yolasite.com/?fbclid=IwAR1qn8g9pvtbvA0imt4helEpAbfSFW6AM-_rsF6WIj6D2SCOQxskTZB_Wew.

8.1 "*Following in Greta's Footsteps... Towards the Social Nursery*"

At the beginning of 2019, it was common to find various activities posted on the social network Facebook that showed concrete projects to deal with the climate crisis in Milan. One that caught my eye was promoted by the Cascinet urban agriculture and cultural association. The association had been given the right to use an area located at the beginning of the *Parco Agricolo Sud Milano*. This area was divided between two groups: one was to create a commercial urban vegetable garden (Vettabbia); the other was to dedicate itself to creating an urban agroforestry. Why an agroforest and not just a forest? The aim was to "revive compatible agricultural and forestry practices that existed before the green revolution transformed agricultural practices in Italy and practically everywhere else in the world" (Alessandro, 09/019). Despite the apparent simplicity of the project, it was part of a wide-ranging urban regeneration network promoted by *Milan's* city council—*Milano Porta Verde*[3]—together with different local partners and universities: Agriporto, CasciNet, Algaria, Terzo Paesaggio/Az. Agr. Davide Longoni, FioreUrbano, CURE, Deafal NGO) and the University of Milan and Politecnico di Milano.

The Vettabbia agroforestry project was part of an initiative promoted by OpenAgri, an innovation centre (which includes the Municipality of Milan) whose aim was to develop solutions to link rural and urban agricultural practices in the Milan region.[4] According to Alessandro, one of the main people responsible for the project,

> "We believe that the involvement of citizens in the formation of a conscious and active community in urban agroforestry projects can be a lever for the development of a new civic sense that acts as a link between issues such as food production and consumption, protection, care for the soil and

[3] https://liaison2020.eu/ambassadors/milano-porta-verde/.

[4] The municipality of Milan is taking part in UIA—Urban Innovative Actions, a European Union initiative that promotes innovative actions in cities, together with 15 partners, including universities, associations and companies, with the OpenAgri project. Funded by the European Regional Development Fund, UIA aims to directly involve urban areas in a large-scale experimentation laboratory to identify "innovative" solutions. In Milan, the focus of the project is the creation of new work and development opportunities in the Porta Romana-Vettabbia area.

> the territory, restoring the sense of authorship of the agricultural landscape for citizen-farmers. The agriculture of the future, integrated with ecology, maximises economic and social resilience, as well as providing solutions to combat the effects of climate change" (Alessandro, 5/09/2018—Facebook).

My intention here is not to assess whether these projects have achieved the objectives stated by their promoters or funders. What interested me was observing how they managed to attract different profiles of participants, articulate multiple activities and open up spaces for expressing concerns about the contemporary environmental crisis—at the same time as offering concrete opportunities for experimenting with personal and collective transformations, both material and emotional.

In the case of the climate activists, Alessandro was explicit in saying that the arrival of members linked to the FFF and XR had "revitalised" the camp, giving new relevance to regeneration practices. In addition to the usual activities, the public meetings have become moments of articulation between these regenerative practices and the ecological and climatic situation, both local and global. Agroforestry was presented as a concrete gesture towards another possible city:

> "On Saturday, before we took part in Milan Strikes for the Planet! - Strike for the Climate, we prepared the ground for Agroforestiamo - an act of regeneration. We believe that the regeneration of territories and people is the starting point for mitigating the effects of climate change" (17/03/2019).
>
> "Next meeting on 15 June with our friends from Fridays For Future and Milano per il Clima to understand why agroforestry can mitigate climate change" (15/06/2019).

The marginalisation of the experience—both political and geographical—should not be read as a mere continuation of counter-cultural practices. Rather, it is a consequence of the fossil urbanism assemblage. The spaces activated by similar experiments are always those temporarily forgotten, neglected or abandoned by public authorities and by the logic of property market valuation. In the paradigmatic case of Scalo di San Cristoforo, they went from oblivion to strategic interest, and with it the destruction of a dense network of interspecies relationships that had formed there—as we'll see in our second case study. Even so, as in other territories (Goccia, Baiamonti), small urban fragments that weren't captured by the cement machine have become the nuclei of new collective practices, where more-than-human forms of sociality capable of resisting the territorial homogenisation of the resilient city are being tested.

Activists know they have to be vigilant. The fossil complex lurks on all sides: it's a capture machine. And experimenting with dissident practices does not exempt them from ambiguities. The Vettabbia experience was, for example, incorporated into the Forestami programme and codified as public policy on mitigation and adaptation. While "planting trees to save the planet" may seem like an emancipatory gesture, it doesn't automatically mean freeing oneself from the bonds of everyday fossilised sociality—or escaping the risks of exclusion inherent in the grammars of the Resilient City. It's worth noting, by the way, the scant participation of immigrants or minorities: the majority of those involved were white Italians (with a few Latin American, French and Spanish exceptions).

Participants engaged in regeneration practices even without being able to claim full autonomy from fossil infrastructures. However, these practices can be read as forms of bricolage (Viveiros de Castro 2019), in which technologies inherited from the "green revolution" are re-appropriated and re-signified to respond—precisely—to the consequences of this same paradigm. The subjects inhabit an environment traversed by fossil residues, but this does not imply capitulation or connivance. These are the concrete conditions of their experiences of becoming active, and it is this problematic and ambivalent character that they seek to transform in practice. It is in this environment—modern, obsolete, ruinous—that they construct partial moments of a life that escapes extractive meshes. It is from there that their responses gain meaning and relevance.

My first field visit coincided with the start of the agroforestry project. Part of the initial planting (2019–2021) had already taken place, with hundreds of newly introduced trees. Maintenance was essential so that these plants wouldn't be suffocated by other species. The work that morning would be done using a traditional Lombardy technique: compressing the emerging vegetation with a wooden stick attached to a rope, without cutting it. There were eight of us, most with no previous experience. And so, the activities continued throughout 2019 and 2020, with breaks and adaptations due to the covid-19 pandemic: maintenance, planting, workshops and courses on agroecology and species collection. A diverse group, aged between 18 and 60, made up of students, precarious workers, architects and artists.

The main people in charge of the area, Alessandro and Enrico—both intermittent workers, constantly looking for funding—led the actions with enthusiasm and anguish. Members of the Cascinet association often expressed concern about the precariousness of the project's continuity. Their dedication to their work, with no guarantee of stability, moulded the collective mood. Passionate about Swiss agronomist Ernst Götsch, they often referred to his experience in Brazil—not just the syntropic techniques, but the way those practices involved other ways of life, reorganised manual labour and reconfigured the political subject. References to the MST, for example, indicated an interest in alternative forms of socio-economic and territorial organisation, outside the capitalist production logic. Syntropic agriculture, by uniting living beings in a process of mutual complexification, offered both new possibilities for association and new ethical and political demands.

According to Ernst Gotsch,

> "At Syntropic Farming, we work from designing the arrangements with different species, through implementation and then continuing with each stage of managing our plantations so that they produce their own fertiliser. To do this, we plant high-density trees, grasses and herbs that have in common the characteristic of growing easily and vigorously after pruning. And you manage them accordingly. The effect of this pruning, which is done periodically, results—in addition to increasing the light supply for our crops—in organic matter in large quantities which, when put into the soil, creates thriving life in it and, indirectly, fertiliser for our plants".[5]

Developing these perspectives in an urban area represented a significant technical challenge, but the results obtained over two years show that it is feasible. Accompanying forest succession meant, for those who didn't master the technical details, observing up close how different temporalities and forms of life slowly emerged in a territory that used to be a rubbish dump. According to Carlo (22/05/2020), a participant in the activities: "It changes you. You start to pay attention to these differences and follow the variety of life that can emerge". As noted in the introduction to this book, becoming active does not correspond to a sudden realisation of a previously ignored situation, nor to the adoption of a fixed

[5] https://agendagotsch.com/pt/diferencas-entre-A-agricultura-sintropica-E-organica-2/

identity—that of a "true activist", distinct from the rest. It is perhaps more appropriate to understand this dynamic from Knox's notion of thinking like a climate: the continuous act of making explicit the material links and effects that run through everyday experience. These subjects refer to specific situations that have become problematic—and in doing so, they interact with objects, technologies and fossil products that impose themselves as actively communicative materialities. These elements emit signals that can be interpreted, both in terms of their conditions of production and their local effects on a specific ecology and climate. This is an environment whose interactivity dynamic interferes in the constitution and reproduction of different modes of existence.

Starting with an apparently simple statement—"plastic packaging must be banned"—made during an activity in the Vettabbia camp, the participants began to follow the life cycle of these objects and the structural conditions that make their production possible. As we saw in the case of Block Friday, this investigation leads to the identification of production chains and the questioning of the desires that sustain consumption in the Anthropocene (Chakrabarty 2020). Consumption, desire and even corporeality itself are now seen in the light of the exploitation of natural resources and their contribution to climate change. But once again, it's not just a question of raising awareness. Doubts, moral tensions and feelings of guilt run through these experiences, with emotional, political and ethical implications. It's about redirecting flows, blocking circulations, refusing captures—and, at the same time, regenerating, transforming oneself and the environment through a new relationship with more-than-human forms of life.

From mid-2019—even with the pandemic restrictions in 2020—many who considered themselves climate activists began to take part in activities. At the same time, individuals with no ties to formal organisations did not cease to recognise themselves as activists: what qualified them was doing something or failing to do something. Becoming active positioned them as agents capable of interfering in ecological and climatic conditions, and in the material devices that organise the reproduction of everyday life. For A. (XR), it was necessary to reactive historical experiences that showed that "other cities" were possible. She recalled Paris in the nineteenth century:

> "There were thousands of gardens. This meant a relationship with the land, which was preserved in quality and guaranteed food. We need to realise

that human beings are part of nature, that they are animals. The idea-spread with Covid - that humans are a virus is destructive: humans can also take care of, collaborate with, help nature" (18/06/2020).

The question that ran through these experiences in a practical, emotional and reflective way could be formulated like this: How can I do something to slowdown the climate crisis that is underway? No action was considered trivial: everything required reflection. Preparing a pizza, for example, became an object of negotiation—vegan and vegetarian diets had to be respected. The pizza oven, built with raw earth, was almost entirely made from local materials. The idea was to demonstrate in practice that it is possible to solve every day needs with resources available in one's own territory. The participants were visibly impressed. The work was hard: 15 lorries of earth excavated in the park itself, transporting materials, mixing clay with sand and straw. At least 20 people were involved. The atmosphere was festive, despite the health restrictions.

For Anna, "it was marvellous to be with other people who were trying to do something about the climate, even without knowing exactly how to do it". According to her, this opened "a new horizon of possibilities". Mercedes, a human rights activist and musician, brought a solar oven to demonstrate that it was possible to cook without relying on fossil fuels. Curiosity was visible. People approached, asked, touched and speculated. The pedagogical gesture, in this context, was also a political—and speculative—gesture: what would it be like to live without the continuous use of fossil fuels.

By defining the Vettabbia space as an experimental environment, my aim is to emphasise that nothing has been definitively established there. It's a process of learning—of acquiring technical knowledge that makes it possible to exercise autonomy beyond fossil materiality, while constituting an open relational space. What can be done in a global metropolis like Milan? The soil, contaminated by heavy metals, requires specific care; the tools, inherited from a previous generation, are often of little practical use. Those who work daily do so, for the most part, without pay—which is not to say that they do so by voluntary choice. The people who gather there, each in their own way, often must start from scratch, which requires time, patience and a kind of persistent commitment to the work required.

It is a space of ruins that, paradoxically, presents itself as a field where it is possible to learn how to live in a future of crises. And it is from this affective and political experience that the subjects decide to participate. Regenerating becomes a gesture of hope—ephemeral and limited, perhaps, but powerful enough to produce a transformation in the way these subjects situate themselves in relation to the fossil environment. It's not a question of naivety, but of learning to live among ruins, with the knowledge that new catastrophes will surely come. It's not an irrational hope in the arrival of a definitive solution, nor is it paralysing pessimism. It is, as Haraway (2016) proposes, staying with the problem. What matters is how this "staying with the problem" will be done.

The distinction between chains of dependence and relations of interdependence offers two crucial categories for grasping the ambivalences that run through the ways of inhabiting urban spaces dominated by fossil infrastructures. They imply a situated and relational inscription; their concrete forms vary according to the environment and the composition of the agents—human and non-human—involved. The ambiguity—of practices and discourses—is not in the subjects, but in the material configurations in which they are inscribed, in which multiple agencies intertwine and reconfigure themselves.

And it's not necessary to look at distant examples to see how these tensions operate. Even in an apparently restricted context like Milan, changing your diet or abandoning your car cannot be considered a universally applicable gesture. What alternatives are there for making the same journey without a car in a reasonable amount of time? How can we change our eating habits without incurring higher costs, especially compared to the prices offered by the big supermarket chains? We don't have a definitive answer, but it's essential to recognise that this type of questioning is present in the activist debate—as witnessed at the table on urbanism and climate during the first edition of Klimat Fest in 2020. As one participant remarked: "How can you give up your car living in the countryside if there's no public transport to get around independently?".

The lack of infrastructure capable of sustaining ways of life that are not linked to fossil fuels—especially in the spheres of food and consumption—has led these collectives to value small urban and community gardening experiments. This recurrence is evident in the events organised around food and the climate crisis. The return to vegetable gardens seems to indicate, more generally, the importance of looking at the interstices: it is in these gaps—marginal and provisional—that activists seek to constitute dissident practices, even if limited and ephemeral, in the face of business as usual.

As I have already discussed, it is a question of considering the responses in their materiality, without neglecting the practical and differential effects they produce—since modifying socio-material assemblages implies new bodily and subjective dispositions. In the urban context, this means gaining access or being excluded (Easterling 2021); it is in this process that forms of climate citizenship emerge, articulated with new governance regimes on local and extra-local scales. The strong presence of the food issue in Milan cannot be dissociated from the city's institutional commitment to formulating a public food policy—one of the consequences of Expo 2015. But, as we shall see, this process cannot be reduced to a direct result of the Expo either: there is a longer and more diffuse history underpinning these practices.[6]

The gardens in the Milan metropolitan area total more than 2000 units.[7] This means considering a variety of ways of using space in a very large area. The classification gives by urbanists differentiates the types of gardens into two: those in which horticultural and productive practices prevail (predominant throughout the territory of Milan, mainly in the south of the city and in the metropolitan area); and closed colonies (located mainly in the north of the Municipality of Milan and in the metropolitan area), defined by a very heterogeneous set of private or privatised open spaces, associated with different practices and methods of use and life not necessarily linked to cultivation. Most of the gardens (including the occupied area) are fenced gardens. Another differentiation refers to management: between those managed by associations, organisations and town halls and those managed by individuals and informal groups. The predominance of gardeners is male. In private gardens, the role of women is generally linked to support activities, care and organising group meetings (parties, get-togethers with friends). In the case of gardens managed by associations/municipalities, the role of women is more significant, but smaller in any case. The predominance of the

[6] As stated in the Climate-Air Plan, "With regard to the municipal reality, for the city of Milan, the theme of agriculture is of notable importance in relation to Food Policy, sustainable territorial planning and urban afforestation. It is interesting to note that the city of Milan is the second largest agricultural city in Italy, with 2,900 hectares of cultivated agricultural land (ISTAT), a figure that demonstrates the relevance of the issue for the Milanese territorial system and the adaptation measures to be taken (ACP: 786)".

[7] Our date came from the research carried out by Mario Cucchi, Daniela Gambino and Antonio Longo. *La citta degli orti*. coltivare e costruire socialità nei piccoli spazi verdi della grande Milano. Macerata: Quolibert, 2020.

over-60 age group is a result of both the availability of time and a public policy aimed at prioritising the availability of services for this age group. So, there is nothing naturalised (the vegetable garden as an activity for the elderly). About technical competence, the vast majority are self-taught (family tradition; working directly on the land and learning through group work).[8]

As already discussed in Chapter 3, the explicitness of the climatic conditions and the chains of dependence lead these subjects to a type of empirical research in which it is a question of following the matter, its connections and its *assembly*, and whose relevance can be followed by the continuous training workshops. They will encounter other agents and other stories, as well as different experts, models and scientific theories: and it will be in this process that the inhabited area will become critical, full of hitherto unknown activities, active to the point of interfering and changing habits and practices, perspectives and bonds.

8.2 "We Need to Get to the Duomo": Agroforestry as a Peripheral Strategy

> "A hectare where peaches, apples, cherries, figs, plums and more than 2,500 other small fruit trees coexist. Between the rows of vines, gooseberries, Jost berries, raspberries and other lower plants grow in harmony with the local vegetation."[9]

In the Corvetto neighbourhood, a southern suburb of Milan at the start of *Parco Agricola Sud*, a project was launched in 2018 to create an agroforestry system less than 15 minutes from the city centre. Born from the granting of the use of the space by the city of Milan to Cascinet, a social promotion and urban agriculture cooperative, this project has

[8] 80 per cent of the places are offered by the public administration. Two resolutions (1143/2012 and 451/2015) seek to frame the creation of new gardens from a multi-functional perspective. This is precisely in order to open them up to other uses in the public interest and to make gardens innovative spaces for new practices of living in and inhabiting the city.

[9] https://www.ilfattoquotidiano.it/2021/12/14/milano-cosi-I-campi-abbandonati-del-parco-vettabbia-sono-stati-recuperati-con-laiuto-dei-residenti-ricreato-lecosistema-di-un-bosco/6421940/?fbclid=IwAR2GT31BG6pGmorGXPE3aPpAaH2PW_Q5r07Ch_d3tRq984ykybm0AhYBKvc

developed a peculiar practical and discursive repertoire to deal with the ecological and climate crisis in urban spaces.

The creation of the agroforest had two main objectives: to increase the forest cover at Milan's southern gateway and to produce varied and healthy food[10] through an approach that involved building a collective of practitioners; the aim was to involve the residents of the neighbourhood and transform the urban space which, according to the residents, was abandoned.

According to Alessandro, one of the main promoters of the experiment, "agroforestry represents a low-cost solution for purifying the air in cities and combating the effects of climate change" (12 September 2018). Preparatory activities began in 2018, initially with members and sympathisers of the group. In 2019, planting began, with around 10 to 15 people taking part. As climate mobilisations progressed in different cities, membership grew and diversified, attracting local activists in particular. Meetings with activists from Fridays for Future and Extinction Rebellion multiplied—the latter regularly uses Vettabbia's space for meetings and to organise actions.

The activities carried out can be grouped into three main areas: (1) planting and maintaining trees; (2) courses on collecting, selecting and planting seeds; and (3) expanding the infrastructure through self-construction workshops. Four workshops were held during the research period: the construction of a shelter to store tools and offer protection from the weather; the building of a wooden bridge linking two areas separated by a watercourse; the construction of a raw earth oven with local materials—which involved four days of intense and highly participatory work, with an emphasis on the presence of young people and women—and, finally, a workshop to create a bio-lake, inspired by permaculture techniques.

These practices and knowledge were presented by Pietro (a member of the Cascinet association) as sustainable devices, capable of giving the participants a certain autonomy from the carbon-intensive devices. The degraded and abandoned space is thus re-appropriated through a broader regenerative logic—not by calculating how much CO_2 will be sequestered or how much food production will be generated, but as an affective experience of an attempt to desert fossil infrastructures. As C. (a member of

[10] As one of the concessions was for food production, Vitabia is trying to recover the money it invested because it didn't know the area was unsuitable for food production.

XR and an active participant in Vettabbia) said, it was necessary to "learn to live within the climate catastrophe": learning how to make an oven, set up a vegetable garden, build a bio-lake, would be ways of acquiring useful technical skills.

The difficulty of planting and maintaining the trees "taught me that I had to be patient", said Andrea (38). What's more, "I had to learn the resistance inscribed in the trajectories of the soil and the plants". For him, it was essential to "synchronise the rhythms, without trying to impose oneself violently"—something that is reminiscent of what Tim Ingold (2019) calls learning by doing, a social and subjective experiment of involvement in the art of (re)generating relationships of interdependence.

Inherited technologies—such as brushcutters from the 1960s and remnants of Lombardy's agricultural modernisation—were reused in new ways. No longer to sustain production rhythms that they could no longer fulfil, but to experiment with ways of doing and being in the territory. The practices involved a performative articulation between body, tool and soil: digging, choosing species, planting seeds, monitoring growth. It was necessary to absorb the resistances of non-humanity, to adjust to the specific conditions of the soil, humidity, temperature and the presence of bees—aspects generally ignored by the extractive logic of industrial agriculture.

During a meeting dedicated to evaluating plant growth—which soon turned into a debate on how to "transform the city as a whole"—one participant suggested, between serious and joking, the idea of launching a "decentralised agroforestry guerrilla action": learning the techniques and expanding autonomous actions to occupy abandoned land in the city until it reached the urban centre, "reaching the Duomo". The provocation, made in a relaxed tone, aroused interest because it signalled a desire to articulate regeneration practices with forms of direct territorial resistance. The proposal took on speculative and antagonistic contours: literally breaking up the urban fabric with the "blow of a hoe".

Regardless of the intention to put it into practice, the most significant thing was the type of problem the situation posed. The paradoxical condition of the local scale—at once a point of mobilisation and a limited space for intervention—revealed the fundamental dilemmas faced by those who wish to confront fossil urbanism. Local action is necessary, but insufficient if it is not articulated with extra-local dynamics. At the end of the chapter, I will propose a reading that avoids both celebratory optimism and paralysing scepticism: restoring the importance of localised action, without ignoring its limits.

At that moment, a kind of consensus emerged among the activist's present—including XR members—that civil disobedience strategies could be effective in destabilising the hegemony of the fossil complex. This possibility was still open, but the ongoing experiment, rooted in the materiality of the soil and in the collective imagination, seemed to offer concrete clues to other ways of acting, thinking and living on a planet in transformation.

The meetings in Vettabbia Park became privileged occasions for the circulation of knowledge and devices that could be incorporated into everyday life. The activists engaged in a politics of proximity, demonstration, contagion and imitation. This policy involved considering the links between body and materiality based on the emotional and political effort invested in trying to incorporate alternatives. As Enrico remarked, "knowledge didn't just pass through the head, but was learnt from nature, by touching it": knowledge had to be incorporated so that it could guide people within complex socio-ecological systems.

Manual labour established a direct relationship with an active materiality, which in turn acted on the bodies and affections of the participants. This produced a subjectivity that was sensitive to the emergence of other possible worlds—or, as Ingold (2019) proposes, a subjectivity entangled in the materiality of action and experimentation. The aim was to create conditions for the emergence of more-than-human agencies, and of course, new subjects.

The regeneration practices operated as positive capture devices: although limited in time and space, they engaged the subjects in technical and sensory journeys—planting, caring for, maintaining—capable of activating other relational universes. Perhaps it was this practical, embodied and demonstrative dimension that made it possible to maintain an active collective of between 15 and 40 people per meeting. To show, through concrete examples, visible constructions and direct relationships with materials, objects and non-human living beings, that it was possible to inhabit, produce and live in a different way even in a metropolis like Milan.

Ignoring this kind of micro-engagement in the name of an exclusively macro-structural perspective would be a mistake. Material politics, in this case, offers the subject a point of support from which they can begin to reconfigure their relationship with the world and, in this gesture, imagine other possible worlds. In short, it is a question of configuring a territory that is detached—albeit partially—from the ubiquity of fossilised sociality. As Debaise and Stengers suggest, these experimental practices reverberate with Guattari's notion of existential territory: "Ce que Félix Guattari nommait 'territoire existentiel' peut désormais entrer en résonance avec des pratiques multiples d'apprentissage qui conjuguent résistance active et expérimentation des moyens de régénérer ce qu'on pourrait appeler une culture de l'interdépendance" (Debaise and Stengers 2021: 135–136).

Vettabbia's experience, however, is not without its ambiguities. In the end, it will be inscribed and formally codified in the terms and limits set by the municipal administration. But, as discussed in Chapter 2, this institutional codification does not exhaust the meanings or unilaterally determine the ways of living in the resilient city. More than the project organisers themselves, what brought us to the agricultural fields of southern Milan was the desire to accompany the heterogeneous group of people who, albeit temporarily, had gathered there to explore possibilities for escaping the mechanisms of fossil chains—and thereby cultivate conditions for another form of urban habitability.

8.3 Regeneration Practices with/against the Resilient City

Apprehending the more-than-human worlds, as I observed during my fieldwork, is not equivalent to a sudden unveiling of hidden meanings. It's not about a naked subject facing a world revealed in its fullness. In an increasingly expressive world—i.e. one that actively interferes and reacts to human affairs. Subjects "become sensitive" to the presence and action of non-human agencies through specific devices and relations of interdependence that are rooted in particular territories—and whose trajectories are always imbricated by the effects of chains of dependence.

These encounters, although partial and precarious (or perhaps precisely because of this), become occasions for a narrative elaboration that links lived experiences to the territory, in explicit opposition to the disembodied narratives of modernisation, progress and economic growth. Those who become active must learn to "read" these manifestations, articulating minimal daily actions with the complex ecological and climatic interactions that structure the environments in which they live.

By situating this "becoming active" in the Resilient City, my aim is to create contrasts that allow us to follow the attempts at personal and collective transformation promoted by climate activists. By following their practices, it becomes clear that their actions affect a wide range of spheres—energy, food, mobility, consumption—in other words, all the technical and subjective conditions of social reproduction sustained by fossil infrastructures. This situated problematisation of the material and discursive links that structure the urban environment is central to understanding their practices.

This urban is made up of a materiality that is animated, manipulated, contested and mobilised. It interferes, modifies, "makes-do"; it represents both an obstacle and a vector for divergent trajectories. Activists describe their position in relation to the urban machine as a trap: an architecture of capture that requires constant attention to avoid being swallowed up. The "world", however, can become active in different ways, thus challenging the very idea of a stable context or shared point of view.

In an attempt to escape fossil sociabilities, activists create situations that favour the emergence of interdependent encounters and new forms of association. Welcoming this "animated world" into the urban space implies producing other senses, both individual and collective, that can be shared—what Stengers (2020) calls sense-making in common. It's about reactivating interspecies relations through which non-human agencies gain relevance in territorial production, both discursively and materially.

In Milan, this also implies a contrast—and often a tense coexistence—with official urban afforestation initiatives promoted by public–private partnerships, such as the Forestami project. These initiatives operate within the logic of regeneration as an engine for urban development and economic growth—a logic that characterises so-called climate urbanism.

The case I'll be discussing is the activities of the group initially called *Scalo Rigeneriamo Urbano*, and more recently *L'Olmo in Fiore*. Formed in 2017, the collective arose from the desire to regenerate the San Cristoforo railway yard, an area abandoned for more than 30 years located on the border between Milan and Corsico. The group, made up of young people aged between 18 and 29, decided to occupy the land—owned by the state railway company—to turn it into a collective garden and community space for events and socialising. As Matteo, one of the members of the collective, explained, these practices act as healing processes not only for ecological and climatic disturbances, but also for subjective, social and affective disturbances. Participants seek to acquire a sense of presence in which non-human agentivity is part of a collective becoming that resists fossilised sociabilities, opening up paths to other forms of coexistence and urban transformation. Socio-ecological regeneration, in this sense, emerges as a practical and sensitive response to localised problem situations—and as a way of inhabiting the ongoing collapse without resigning oneself to it.

Talking to Matteo, he told me that the aim of the group occupying that area was to "take care of that place and create a community". In a post on the group's Facebook page, they wrote.

> "We've installed a new notice board at the entrance and repaired the one that already existed inside the oasis, where we can leave traces, messages, ideas for getting to know Scalo better and offer indications about our initiatives. Periodically, we try to clear the area of rubbish and abandoned objects, plastic and dirt; we also did this last 25 April, celebrating Liberation in our own way, because - if you look closely - choosing to care is the exact opposite of the terrible 'I don't care' paradigm, anathema to fascist irresponsibility."

This area, however, has been put up for sale as part of an international competition linked to the C40 network, called Reinventing Cities. The master plan for this area is to turn it into a "natural park" in contiguity with another area of the railway that was demolished—this one to be residential—where a new building designed by Stefano Boeri—Forestami's scientific director—called "*Bosco orizzontale*" should appear by the end of 2022, with the price per square metre already estimated at €6,000. In other words, self-defined regeneration projects that operate from very different perspectives and with obviously different practical consequences. I can't discuss both in detail; I can only emphasise that the presence of plants in the city cannot be evaluated in the abstract, but rather that from each experience it is possible to extract the modes of existence that these plants can acquire, where they will be confined and what potential they will stimulate or block.

The first time I visited the site, in 2019, the members of the collective were preparing the ground for a flower planting event. The dynamics of the occupation were reminiscent of other experiences I had followed in the city, such as Vettabbia, which the participants themselves cited as an inspiration (Matteo had actively participated in that process). The first challenge was to clean up the site, which had become a rubbish dump after more than 30 years of neglect. But this clean-up couldn't mean indiscriminate disposal: the spontaneous ecology that had emerged during this period had to be preserved.

Inspired by Gilles Clément's notion of the third landscape (2015: 55), defined as "all the residual landscapes resulting from the organisation of spaces and omitted by it", the activists sought to suspend the naturalistic judgements that relegate these marginal ecologies to a secondary role in the multi-species diversification of the territory. Winding paths were laid out to allow transit between the vestiges of an urban modernisation in ruins—in the shadow of an unfinished architectural heritage building, considered to be of great historical and aesthetic value, designed by Aldo Rossi. For the members of the collective, regenerating an urban space doesn't mean pouring concrete, but rather cultivating relationships between people, ideas, culture and plants.

> "To regenerate the fertility of the soil and prepare it for the oasis to come, today in Scalo we are sowing subterranean clover, capable of fixing nitrogen and self-seeding. We've piled up the seedlings left in the ground since spring, discovering a soil that is already darker and more inviting than anywhere else. Here, in irregular shapes and curves, we sow. Spontaneously, grasses like Sorghum Alepense are spreading, and we hope they'll be good companions."

The members' work involved not only caring for the land, but also mediating between the territory and the residents of the neighbourhood. Some residents saw the occupation of the land as a way of preventing "situations of social degradation", such as drug trafficking. There were latent tensions: inside Aldo Rossi's unfinished building, a growing population of immigrants caused fear among the residents. One Sunday morning, a man walking an imposing rottweiler commented, with racist irony, that the "dog was black because it only ate black people". Faced with my shocked expression, Matteo—a member of the collective and openly anti-fascist—commented that, as I could see, his work involved much more than plants: it also required confronting the local social ecology and its contradictions.

The members of the collective maintained, according to them, a cordial relationship with the migrants who occupied the building and tried to think of regeneration as a political gesture that connected ecology, urbanism and coexistence. Regeneration of space, they said, could not be understood only in ecological or technical terms, but involved the production of new ways of living, amid the tension between migration policies, urban imaginaries and territorial inequality.

> "Here we are, ready to take on the responsibility of planting new small trees in Scalo and sowing beneficial grasses for the soil. Come along on Wednesday 14 November. We'll be here from 3 pm until sunset - it'll be a form of sharing, of common action. Or stay at home and enjoy the GDP, worry about the virus, be afraid of the migrants - surely that's more important than the fertility of the soil." (Facebook post).

In the following months, the collective decided to start a dialogue with the owners of the area—*Ferrovie dello Stato* (FS)—in search of a negotiated solution. However, after an initial cordial response, FS broke off talks and decided to close off the regenerated area, turning it into a dumping ground for materials for the future redevelopment of the railway yard as part of the Reinventing Cities competition promoted by the C40 network. Along the railway towards the centre of Milan, the forest now codified in the language of climate urbanism has been promoted as a technical device for urban-climate regulation—an operation that masks the conflicts and social erasures that accompany this type of planned regeneration. Although Boeri's project was not located in the occupied area, it was part of the neighbourhood's green redevelopment project and gave rise to countless pages dedicated to yet another real estate development with "great potential to promote a ripple effect throughout the area" and attract more investment:

> "The façades will also have lush vegetation: 170 trees and more than 8,000 shrubs. The plants and trees, however, will not just be decorations, but a kind of green cooling system: "Climate change will increase the number of hot days in cities in summer, which is why for several years we have been working on a new architecture in which vegetation is not a decorative element, but a structural one," explained Stefano Boeri during the presentation. According to a study cited by the architect, the foliage of the trees will help reduce solar radiation on the structure, just as it does in the "Bosco": "A study of the buildings in Porta Nuova revealed that the leaves of the trees filter out most of the sun's rays, which allows for a reduction in heat on the façades of around 30 degrees. In Bosco Verticale, air conditioning is practically not used during the summer because the plants' cooling system is efficient."

Trees and shrubs are introduced as technological components in architectural and urban planning projects, making up a functional green infrastructure. The very notion of infrastructure, as we know, raises numerous questions that have already been widely debated in the field of the anthropology of infrastructure and science and technology studies. However, when the forest is mobilised as a solution to urban climatic and ecological problems, it also opens up space for perspectives that tension and challenge current urban policies. As in so many other experiences in the city, the tree becomes an ally in the co-production of subjectivities and forms of collective action that escape the logic of mere functionalisation. Instead, it points to the production of multi-species spaces—albeit ephemeral and precarious.

> "Regeneration also begins in the soil, in its balance, in its silent and proactive microbial population. It would be easy to buy bags of soil, import an 'allogenic' product, enrich Scalo with ready-made resources from abroad. Whenever possible, we prefer another way: subversive, stubborn and proudly inefficient for our time. We chose to recover raw organic matter - organic waste, vegetable waste, ashes from the neighbourhood pizzeria, wood scraps - to produce a new compost from the surrounding neighbourhoods themselves. This urban path of collecting waste and converting it into resources is slowly consolidating itself as a virtuous circuit, activated by the community synergies that surround us."

In this chapter, I've tried to explore some aspects of what it means to become active in an active world. In the end, the central point I return to is the question of the context of action. As we have seen, "context" remains fundamentally open: there is no question of presupposing a set of general empirical data that delimits in advance a relevant universe for experience—in other words, you can't resort to an external, totalising point of view from which to judge the relevance of actions. The context is disputed, and the dispute revolves around what should or shouldn't count for the subjects who experience their activation: becoming active is, in itself, a more-than-human experiential process; one doesn't respond to a predefined problem, but rather configures the very problem that demands an answer.

The question of how to respond to climate change is immanent to the situations and is faced by those who are entangled in its consequences—with no guarantees of the justice or effectiveness of the possible responses. Because they are situated experiments—in other words, territorially rooted and spatially localised—these practices must be understood as vectors for producing responses to the climate crisis and ecological collapse. Faced with a dominant paradigm that proposes technical and systemic solutions to the climate, it is essential to turn to experiments that test other forms of association. Forestry, in this sense, acquires a political-affective dimension: it becomes a means of manufacturing low-carbon socialities, and not just regulating ecosystem services.

When fossil fuels are recognised as agents of the climate crisis, adaptation and mitigation are not limited to protecting urban materiality or the technocratic energy transition. They imply the continuous infrastructuraction (Nolan 2018) of dissident relational spaces. If plants gain opportunities for multi-species association in urban spaces, humans, in turn, are invited to open up new paths that destabilise the continuity of ways of life based on extractive chains and fossil dependency. Small, but not insignificant.

References

Chakrabarty, Dipesh. (2020) When the global reveals the planetary: Bruno Latour interviews Dipesh Chakrabarty. In: Latour, Bruno; Weibel, Peter (eds.). *Critical zones*: the science and politics of landing on Earth. Cambridge and Karlsruhe: MIT/ZKM.

Clement, Gilles (2015). *Piccola pedagogia dell'erba*: riflessione sul giardino planetario. Roma: Derive Approdi.

Debaise, Didier; Stengers, Isabelle. (2021) Résister à l'amincissement du monde. *Multitudes*; 85: 129-137.

Deleuze, Gilles and Guattari, Felix. (1987) *A thousand plateaus*: capitalism and schizophrenia. Minneapolis: University of Minnesota Press.

Easterling, Keller. (2021) *Medium design*: knowing how to work on the world. London: Verso.

Haraway, Donna. (2016) *Staying with the trouble*: making kin in the Chthulucene. Durham: Duke University Press.

Ingold, Tim. (2019) *Making*: Antropologia, archeologia, arte e architettura. Milano: Raffaello Cortina Editore.

Kohn, Eduardo. (2021) Come pensano le foreste: Per un'antropologia oltre l'umano. Translated by Alessandro Lucera and Alessandro Palmieri. Preface by Emanuele Coccia. Milano: Nottetempo.

Latour, Bruno. (2018) *Down to earth.* politics in the new climate regime. Cambridge: Polity Press.

Marres, Noortje. (2015) *Material participation*: technology, the environment and everyday publics. London: Palgrave Macmillan.
Sassen, Saskia. (2010) "Cities are at the Center of Our Environmental Future." *Sapiens* 2 (3): 1–8.
Schlosberg, David, and Romand Coles. (2016) "The New Environmentalism of Everyday Life: Sustainability, Material Flows and Movements." *Contemporary Political Theory* 15 (2): 160–181. https://doi.org/10.1057/cpt.2015.34
Stengers, Isabelle. (2011) *Thinking with Whitehead*: a free and wild creation of concepts. Cambridge: Harvard University Press.
Stengers, Isabelle. (2020a) *Réactiver le sens commun*: lecture de Whitehead en temps de débâcle. Paris: Les empêcheurs de penser en rond.
Viveiros de Castro, Eduardo. (2019) On Models and Examples Engineers and Bricoleurs in the Anthropocene. *Current Anthropology*; 60:S20: S296–S308

CHAPTER 9

Final Remarks

"This mean, more technically, that is was indeed the future that was to be infected" (Stengers 2011: 419).

In this work, I sought to develop an empirical and speculative investigation into the process of becoming active in the face of climate change in the city of Milan. The results presented here should be understood for both their potential and their limits. As with any overflight, you are travelling over a wide landscape, but it is not always possible to land at each of its points with the same level of detail. To do otherwise—to fixate on a single section—would entail the opposite risk: losing sight of what is articulated around it. The aim, therefore, was to explore connections and fractures, making them resonate and recording the practical consequences of those who seek to think with—and from—the climate crises.

The health emergency caused by Covid-19 has decisively determined the shape of this work. The restrictions on movement and the obstacles imposed by the lockdown prevented the development of the original project. The methodological solution was to turn a limitation into an opportunity by adopting a localised point of view that prioritised monitoring local responses to the climate crisis.

The notion of "response" proved to be a central analytical operator. It allowed me to bring together heterogeneous experiences in the city of Milan and treat them as part of a common problematic, even if not unified. With this, the idea of "becoming active" proved to be particularly

D. Delatin Rodrigues, *Climate Change and Political Action in Italy*, https://doi.org/10.1007/978-3-031-98771-7_9

useful—not only because it avoids the normative disputes around what is or isn't activism, but because it allows us to observe its pragmatic dimension: its subjective, material and relational effects. In addition, this notion broadened the empirical field of research, allowing forms of engagement that are often invisible to be considered relevant, if they were meaningful to the subjects themselves.

The second step was to locate these responses in the socio-material fabrics of Milan's urban space. Given the impossibility of a comparative analysis between cities, I chose to follow diversity from proximity. The aim was to show that the response to climate change is not organised as a universal project, but as a multiplication of localised problems and solutions. Technologies, objects, infrastructures, living beings, temperatures and atmospheres are all part of these responses—which involve both maintaining the conditions for reproducing fossil hegemony and challenging it.

As these chains of dependency jeopardise the Earth's ecological conditions, some people are beginning to build alternatives guided by relationships of interdependence. It is these subjects—agents of regeneration—who seek to recompose territories and open up space for practices that welcome more-than-human agencies. In this process, politics becomes a material practice, orientated towards the reinvention of living conditions—beyond security or human consumption. Subtracting territories from extractivist logic thus becomes a fundamental political gesture.

I don't intend to argue that these practices are enough to contain the ecological crisis. I just try, as suggested by Debaise and Stengers (2022), to follow the modes of composition, intra-action and agency that emerge in the interstices. I know that, given the seriousness of the climate crisis, this approach may seem modest, even insufficient. But threats alone are no justification for turning research into judgement. What we can extract from this adventure, which was also mine, is that those who become active are learning to "make room" (Latour 2013) for peoples, objects and beings that have historically been excluded from modern politics. And this is done in a situated way, through concrete, partial and vulnerable gestures. These emerging geographies, composed in precarious environments, deserve to be told. In them, not only the future of the climate is at stake, but the very meaning of what can still be called collective life.

References

Debaise, Didier, and Isabelle Stengers. (2022) "An Ecology of Trust? Consenting to a Pluralist Universe." *The Sociological Review* 70(2): 402–415. https://doi.org/10.1177/00380261221084794

Latour, Bruno. (2013) *Inquiry into the modes of existence*: an anthropology of the moderns. Cambridge: Harvard University Press.

Stengers, Isabelle. (2011) *Thinking with Whitehead*: a free and wild creation of concepts. Cambridge: Harvard University Press.

Appendix I

Milano Per Clima (M4C)

> AISA Milan-Bicocca section	Scientific portal on environmental issues
> Angeli dei Navigli	Navigli Canoeists Association
> Ridef Area (27/09)	Renewable Energy Association (Polimi)
> Associazione Giacimenti Urbani	Sustainable living
> Bambini and Natura	Cultural association: reflections and experiences on the relationship between human beings, and children in particular, and nature
> Cascina Cuccagna	Lifestyles and everyday practices, the vital relationship between the city and the countryside
> Tourist Centre	Education and training project
> Chiesa Pastafariana Italiana	Religious freedom
> ClimateEngage	Sustainable lifestyle podcast
> Stefano Rodotà Committee	People's Committee for the Defence of Public and Common Goods
> Logistics Coordination South Milan	Defence of the land south of Milan
> Coordination at Sonda	
> CORE-lab	Human resources consultancy and development
> Diciassette	It deals with territory
> ECOREBEL	Sustainable lifestyle
> EGEA Milano	European Geographical Association for students and young geographers
> Fonc	Art collective

(continued)

D. Delatin Rodrigues, *Climate Change and Political Action in Italy*,
https://doi.org/10.1007/978-3-031-98771-7

(continued)

> Fridays for the future—Milan	Climate movement
> Local Greenpeace group in Milan	Environmental NGO
> Lato B	Independent space for socialising, music and culture
> Legambiente Lombardy	Environmental NGO
> LIPU Milan Section	Environmental NGO
> Mamme No Smog Sud Milano	Association for the Protection of Air Quality
> Mani Thesis Milano	Combating poverty and social exclusion and promoting sustainable local development
> Milano Bicycle Coalition	Defence of bicycle mobility
> Milano Cop26	Group for COP-26
> MSOI Milano	Spreading the culture of internationalism among young people and students
> Orto Comune Niguarda	Community garden
> Progetto Mirasole Impresa Sociale	Social housing activities; reintegration into work and cultural improvement
> RECUP	Association working on the issue of food waste and social exclusion
> Repubblica Nomade	Lifestyles: consciousness and nomadism
> Resilient GAP	Socio-environmental activism (Polimi)
> Rete della Conoscenza Milano	Education network
> Seigas Acquisto Solidale Group	Food and the neighbourhood market
> SITmilano	Social innovation
> Status with zero impact	University collective: adaptation to mitigation (Statale)
> TempoZero—Climate Change	University collective (Statale)
> Effective Altruism	"Group of altruists in constant search of how to 'do better'"
> ATLAS Onlus	Networking to think globally and act locally
> Banda degli Ottoni a Scoppio	Art collective
> Fondazione Casa della carità "A. Abriani" Foundation	Social and cultural basis
> Cittadini for the Aria onlus	Air quality
> CEMP Consultancy	Sexual and gender rights
> CrowdForest	Urban forestry and environmental problems
> Earth Strike Italia	Youth collective
> Edizioni Sonda	Publisher (nature)
> Environmental and cultural education AEA Demetra	Innovation and green technologies
> Energetica—social promotion association	Sustainable and social association

(continued)

(continued)

> It's ours	Co-operative supplier of ethical and sustainable electricity created and managed by the user community
> FIAB Ciclobby Onlus	Promotes the use of bicycles for everyday and leisure mobility
> GLAMourga Macau	Art collective
> Il Cinema e i Diritti—Events and analyses	Cineforum
> Il Fatto Alimentare	Food journalism website
> Isola Solidale	Facility for homeless ex-prisoners
> Collettivo Zam	Collective of activists
> La locomotiva di Momo nido scuola d'infanzia	Nursery school
> The asylum	Nursery school
> Leal Lega Antivivisezionista	Animal rights
> Marmocchi pasta—Milan	Milanese citizens organise to take their children to school by bike
> Milano Climate Save	Milano Climate Save is committed to fighting climate change by combating the environmental impact of livestock farming and fishing, promoting reforestation and phasing out fossil fuels
> Piazza d'armi Le Giardiniere Park, Milan	Food consumption, mobility and soil
> Razzismo Brutta Storia	Anti-racist association
> Rete Clima	Circular economy
> Women's March—Milan	Feminist collective
> It is	Research and education centre
> Deafal	Sustainable agriculture
> Naturiamo	Nursery school, children and the rights of nature
> Cittadini Consapevoli	Environmental education
> Fantastic Cycle Tracks and How to Trot	Pro-bike collective
> WWF Lombardy	Environmental NGO
> Italian Climate Network	Centre for research and advocacy on climate change

Bibliography

Amin, Ash and Thrift, Nigel. (2005) *Città*. Ripensare la dimensione urbana. Bologna: Mulino

Amin, Ash and Thrift, Nigel. (2017) *Seeing like a city*. Cambridge: Polity Press.

Anand, Nikhil; Appel, Hannah; Gupta, Akhil. (2018) Introduction: Temporality, Politics, and the Promise of Infrastructure. In: Anand, Nikhil; Appel, Hannah; Gupta, Akhil (eds.). *The promise of infrastructure*. Durham and London: Duke University Press.

Angelo, Hillary. (2019a) Added value? Denaturalizing the "good" of urban greening. *Geography Compass*;13:e12459

Angelo, Hillary. (2019b) The greening imaginary: urbanized nature in Germany's Ruhr region. *Theory and Society*; 48: 645-669.

Baer, Hans and Singer, Merrill. (2014) *The anthropology of climate change*: an integrated critical perspective. London: Routledge.

Battaglia, Debbora. (2017) "Aeroponic Gardens and Their Magic: Plants/Persons/Ethics in Suspension." *History and Anthropology* 28 (3): 263–292. https://doi.org/10.1080/02757206.2017.1289930

Banoub, Daniel, and Sarah J. Martin. (2020) "Storing Value: The Infrastructural Ecologies of Commodity Storage." *Environment and Planning D: Society and Space* 38(6): 1101–1119. https://doi.org/10.1177/0263775820911003

Barry, John. (2012) The Politics of Actually Existing Unsustainability. Oxford: Oxford University Press.

Beck, Ulrich. (2011) *Conditio humana*: il rischio nell'età globale. Roma-Bari: Laterza.

D. Delatin Rodrigues, *Climate Change and Political Action in Italy*, https://doi.org/10.1007/978-3-031-98771-7

Beveridge, Ross and Koch, Philippe. (2019) Urban everyday politics: politicising practices and the transformation of the here and now. *Environment and Planning D: Society and Space*, Vol. 37(1) 142–157.

Biehl, J. and Locke, P. (2010) 'Deleuze and the anthropology of becoming', *Current Anthropology*, 51(3): 317–351.

Brantley, Susan L. (2020). The critical zone paradigm: a personal view. In: Latour, Bruno; Weibel, Peter (eds.). *Critical zones*: the science and politics of landing on Earth. Cambridge and Karlsruhe: MIT/ZKM.

Blok, Anders. (2014) Worlding cities through their climate projects?, *City*, 18(3), pp. 269-286. https://doi.org/10.1080/13604813.2014.908514

Blok, Anders; Farias, Ignacio (eds.). (2016) *Urban cosmopolitics*: agencements, assemblies, atmospheres. London and New York: Routledge.

Blok, Anders; Meilvang, Marie Leth. (2015) Picturing urban green attachments: civic activists moving between familiar and public engagement in the city. *Sociology*, vol.49(1): 19-37.

Boltanski, Luc and Chiapello, Eve. (2017) *The new spirit of capitalism*. London: Verso.

Bonneuil, Christophe and Fressoz, Jean-Baptiste. (2017) *The shock of the anthropocene*: the Earth, history and us. London: Verso.

Braun, Bruce. (2014) A new urban dispositif? Governing life in an age of climate change. *Environment and Planning D: Society and Space*, vol. 32, pages 49 – 64

Braun, Bruce and Wakefield, Stephanie. (2014) Inhabiting the post-apocalyptic city. *Society & Space*, pp. 01–11

Broto, Vanessa Castán; Westman LK. Ten years after Copenhagen: Reimagining climate change governance in urban areas. *WIREs Clim Change*. 2020;11:e643.

Bulkeley, Harriet, and Broto, Vanesa Castán. (2013) "Government by Experiment? Global Cities and the Governing of Climate Change." *Transactions of the Institute of British Geographers* 38 (3): 361–375. https://doi.org/10.1111/j.1475-5661.2012.00535.x

Bulkeley H and Betsill MM (2013) Revisiting the urban politics of climate change. *Environmental Politics;* 22(1): 136–154.

Callon, M. (1989) Society in the making: the study of technology as a tool for sociological analysis. In Bijker, W.E., Hughes, T.P. and Pinch T.J. (eds), *The Social Construction of Technological Systems*: New Directions in the Sociology and History of Technology, MIT Press, Cambridge.

Callon, M. What does it mean to say that economics is performative?, *CSI Working papers series* (5), Paris, 2006.

Carson, Ashley. (2019) Dirty landscapes: how weediness indexes state disinvestment and global disconnection. In: Hetherington, Kregg (ed.). *Infrastructures, environment, and life in the anthropocene*. Durham: Duke Press University.

Chagnon, Christopher W.; Francesco Durante; Barry K. Gills; Sophia E. Hagolani-Albov; Sanna Hokkanen; Sohvi M. J. Kangasluoma; Heidi Konttinen Markus Kröger; William LaFleur; Ossi Ollinaho and Marketta P. S. Vuola. (2022) "From Extractivism to Global Extractivism: The Evolution of an Organizing Concept". *The Journal of Peasant Studies*.

Chakrabarty, Dipesh. (2009) The climate of history: four theses. *Critical Enquiry* 35, pp. 197–222.

Chakrabarty, Dipesh. (2020) When the global reveals the planetary: Bruno Latour interviews Dipesh Chakrabarty. In: Latour, Bruno; Weibel, Peter (eds.). *Critical zones*: the science and politics of landing on Earth. Cambridge and Karlsruhe: MIT/ZKM.

Charbonnier, Pierre. (2020) *Abondance et liberté*: une histoire environnementale des idées politiques. Paris: La Découverte.

Chu, Eric K. (2017) Urban climate adaptation and the reshaping of state–society relations: The politics of community knowledge and mobilisation in Indore, India. *Urban Studies*, 55(8), 1766–1782.

Clement, Gilles (2015). *Piccola pedagogia dell'erba*: riflessione sul giardino planetario. Roma: Derive Approdi.

Coccia, Emanuele. (2018) *The life of plants*: A Metaphysics of Mixture. Cambridge: Polity Press.

Cohen, Daniel Aldana. (2017) The other low-carbon protagonists: poor people's movements and climate politics in São Paulo. In: Greenberg, Mirian and Lewis, Penny (eds.). *The city is the factory*: new solidarities and spatial strategies in an urban age. Cornell: Cornell University Press.

Comune di Milano. (2019) Piano di Governo del Territorio (PGT) 2030: Documento di Piano, Piano dei Servizi e Piano delle Regole. Approved by City Council on October 14, 2019; effective from February 5, 2020. Milan: Comune di Milano.

Crate, Susan. (2011) Climate and Culture: Anthropology in the Era of Contemporary Climate Change. Annu. Rev. Anthropol. 2011. 40:175–94

Crutzen, Paul. "Geology of Mankind." *Nature* 415 (2002): 23.

Davis, Mike. (2010) Who will build the arc? *New Left Review* 61, jan-feb: pp. 29–46.

Dawson Ashley. (2017) *Extreme Cities*: The Peril and Promise of Urban Life in the Age of Climate Change. New York: Verso.

Debinska, Maria Salomea. (2021) Witnessing from within: hyperobjects and climate activism in Poland. *HAU: Journal of Ethnographic Theory*; 11(2): 445–460.

Debaise, Didier. (2008). The Living and its Environments. *Process Studies*, 37.2: 01-15.

Debaise, Didier. (2017) *Nature as event*: the lure of the possible. Durham and London: Duke University Press.

Debaise, Didier; Stengers, Isabelle. (2021) Résister à l'amincissement du monde. *Multitudes*; 85: 129-137.

Debaise, Didier, and Isabelle Stengers. (2022) "An Ecology of Trust? Consenting to a Pluralist Universe." *The Sociological Review* 70(2): 402–415. https://doi.org/10.1177/00380261221084794

De la Cadena, Marisol (2014). Runa: human but not only. *Hau: Journal of Ethnographic Theory* 4 (2): 253–259.

De la Cadena, Marisol (2015). *Earth beings*: ecologies of practice across Anden worlds. Durham and London: Duke University Press.

Deleuze, Gilles and Guattari, Felix. (1987) *A thousand plateaus*: capitalism and schizophrenia. Minneapolis: University of Minnesota Press.

Despret, Vinciane. (2002) Le emozioni: etnopsicologia dell'autenticità. Milano: Elèuthera.

Despret, Vinciane. (2004) The Body We Care for: Figures of Anthropo-zoo-genesis. *Body & Society*; 10(2-3):111-134.

Dewey, John. (2016) The public and its problem: an essay in political economy. Ohio: Ohio University Press.

Dobraszczyk, Paul. (2017) Sunken cities: climate change, urban futures and the imagination of submergence. *IJURR* Vol. 41 (6), pp. 868-887.

Dowling, Robyn; McGuirk, Pauline and Bulkeley, Harriet. (2018) Governing carbon conduct and subjects: insights from Australian cities. Luque-Ayala, Andrés; Marvin, Simon and Bulkeley, Harriet eds. *Rethinking urban transitions*: politics in the low carbon city. New York: Routledge

Dunlap Riley E. and Robert J. Brulle, eds. (2015) *Climate Change and Society*: Sociological Perspectives. New York: Oxford University Press.

Dutreuil, Sébastien.(2020) Gaia is alive. in: Latour, Bruno; Weibel, Peter (eds.). (2020) *Critical zones*: the science and politics of landing on Earth. Cambridge and Karlsruhe: MIT/ZKM.

Easterling, Keller. (2021) *Medium design*: knowing how to work on the world. London: Verso.

Edwards, Gareth and Bulkeley, Harriet. (2017) Heterotopia and the urban politics of climate change experimentation. *Environment and Planning D: Society and Space* 0(0) 1–20.

Edwards, Paul. (2013). *A vast machine*: computer models, climate data, and the politics of global warming. Cambridge: MIT Press.

Fassin, Didier (2012). *The Humanitarian Reason*: a moral history of the present. University of California Press, Berkeley.

Foucault, Michel. *Storia della sessualità*: l'uso dei piaceri. Milano, Feltrinelli, 2009.

Foot, John. (2001) Milan Since the Miracle: City, Culture and Identity. Oxford: Berg.

Gabrys, Jennifer. (2014) A cosmopolitics of energy: diverging materialities and hesitating practices. *Environment and Planning A*, volume 46, pages 2095 – 2109.

Gabrys, Jennifer. (2020) "Sensing a Moving Planet." In *Critical Zones*: the science and politics of landing on Earth, edited by Bruno Latour and Peter Weibel, 254–263. Cambridge, MA, and Karlsruhe: MIT Press and Zentrum für Kunst und Medien.

Galderisi, A., Limongi, G. & Salata, KD. (2020) Strengths and weaknesses of the 100 Resilient Cities Initiative in Southern Europe: Rome and Athens' experiences. *City Territ Archit* 7, 16.

Gilbert, Scott F. et al. (2012) A symbiotic view of life: We have never been individuals. *The Quarterly Review of Biology;* 87: 325–341.

Goh, Kian. (2015) *A Political Ecology of Design*: Contested Visions of Urban Climate Change Adaptation. (Thesis) Urban and Environmental Planning, Yale University.

Goh, Kian. (2020) Flows in formation: The global-urban networks of climate change adaptation. *Urban Studies*, Vol. 57(11): 2222–2240.

Graham, S. and Marvin, S. (2001) *Splintering urbanism*: networked infrastructures, technological mobilities and the urban condition. London: Routledge.

Graham, Stephen. (2018) *Vertical*: the cities from satellites to bunkers. London: Verso.

Grasso, Marco. (2022) *From big oil to big green*: holding the oil industry to account for the climate crisis. Cambridge: MIT Press.

Grasso, Marco, and Daniel Delatin Rodrigues. (2022) "Disrupting to Decarbonise Socio-Energy Systems: The 'Carbon Transformation Axes' Framework." *Energy Research & Social Science* 90: 102657. https://doi.org/10.1016/j.erss.2022.102657

Guattari, Félix. (2013) "Qu'est-ce que l'écosophie?" Textes (1985–1992) présentés et agencés par Stéphane Nadaud. Paris: Éditions Lignes.

Hage, Ghassan. (2012) Critical anthropological thought and the radical political imaginary today. *Critique of Anthropology*; 32(3): 285-308.

Hallé, Clémence; Milon, Anne-Sophie. (2020): The infinity of the Anthropocene: a (hi)story with a thousand names. In: Latour, Bruno; Weibel, Peter (eds.). *Critical zones*: the science and politics of landing on Earth. Cambridge and Karlsruhe: MIT/ZKM.

Hannerz, Ulf. (1992) *Esplorare la città*: antropologia della vita urbana. Bologna: Il Mulino.

Haraway, Donna. (1988) Situated knowledges: the science question in feminism and the privilege of partial perspective. *Feminist Studies*, Vol. 14, n.3, Aut: pp. 575–599.

Haraway, Donna. (2016) *Staying with the trouble*: making kin in the Chthulucene. Durham: Duke University Press.

Harvey, David. (1990) *La crisi della modernità*: riflessioni sulla origini del presente. Milano: Net.

Harvey, David. (2017) *The ways of the world*. London: Profile Books.

Hetherington, Kregg. (2019) Introduction: keywords of the Anthropocene. In: Hetherington, Kregg (ed.). *Infraestructures, environment, and life in the Anthropocene*. Durham: Duke Press University.

Hawkins, Gay (2011) Packaging water: plastic bottles as market and public devices, *Economy and Society*, 40:4, 534-552.

Hodson, Mike, and Simon Marvin. 2010. "Urbanism in the Anthropocene: Ecological Urbanism or Premium Ecological Enclave?." *City* 14 (3): 298–313.

Hodson, Mike and Marvin, Simon. (2009) Urban ecological security: a new urban paradigm?, *International Journal of Urban and Regional Research*, 33, pp. 193–215.

Hodson, Mike, Marvin, Simon and McMeekin, Andy. (2018) The amenable city-region: the symbolic rise and the relative decline of Greater Manchester's low carbon commitments, 2006–17. In: Luque-Ayala, Andrés; Marvin, Simon and Bulkeley, Harriet eds. *Rethinking urban transitions*: politics in the low carbon city. New York: Routledge.

Holling, C.S. (1973) Resilience and Stability of Ecological Systems; *Annual Review of Ecology and Systematics* 1973 4:1, 1-23.

Ingold, Tim. (2019) *Making*: Antropologia, archeologia, arte e architettura. Milano: Raffaello Cortina Editore.

IPCC, 2018: Summary for Policymakers. In: *Global Warming of 1.5°C*. An IPCC Special Report on the impacts of global warming of 1.5°C above pre-industrial levels and related global greenhouse gas emission pathways, in the context of strengthening the global response to the threat of climate change, sustainable development, and efforts to eradicate poverty. World Meteorological Organization, Geneva, Switzerland, 32 pp.

James, William. (2003) *Essays in radical empiricism*. New York: Dover Publications.

Jasanoff, Sheila. (2004) Ordering knowledge, ordering society, in: Jasanoff, Sheila. *States of knowledge*: the co-production of science and social order. London and New York: Routledge.

Jasanoff Sheila. (2010) "A New Climate for Society," *Theory, Culture & Society*, 27(2-3): 233-253.

Kauffman, Stuart. (2000) *Investigations*. New York: Oxford University Press.

Kirksey, Eben; Schuetze, Craig; Helmreich, Stefan. (2012) Introduction. in Kirsey, Eben (ed.). *Multispecies salon*. Durham and London: Duke University Press.

Knox, Hannah; Huse, Tone. (2015) Political materials: rethinking environment, remaking theory, *Distinktion: Scandinavian Journal of Social Theory*, 16:1, 1–16

Knox, Hannah. (2020) *Thinking like a climate*: governing a city in times of environmental change. Durham and London: Duke University Press.

Kohn, Eduardo. (2021) Come pensano le foreste: Per un'antropologia oltre l'umano. Translated by Alessandro Lucera and Alessandro Palmieri. Preface by Emanuele Coccia. Milano: Nottetempo.

Kopenawa, David; Albert, Bruce. (2013) *The Falling Sky*, Cambridge, MA: Harvard University Press.

Lapoujade, David. (2020a) *William James*: empiricism and pragmatism. Durham and London: Duke University Press.

Lapoujade, David. (2020b) *Le esistenze minori*. Bergamo: Moretti & Vitali editori.

Larkin, Brian. (2013) The politics and poetics of infrastructure. *Annual Review of Anthropology* 42:327–343.

Latour, Bruno. (2004a) Has Critique Run out of Steam? From Matters of Fact to Matters of Concern. *Critical Inquiry*; 30: 225-248.

Latour, Bruno. (2004b) *Politics of nature*: how to bring the sciences into democracy. Cambridge: Harvard University Press.

Latour, Bruno. (2004c). "Whose Cosmos? Which Cosmopolitics? A Comment on Ulrich Beck's Peace Proposal." *Common Knowledge* 10 (3): 450–462.

Latour, Bruno. (2013) *Inquiry into the modes of existence*: an anthropology of the moderns. Cambridge: Harvard University Press.

Latour, Bruno. (2017) *Facing Gaia*: eight lectures on the new climate regime. Cambridge: Polity Press.

Latour, Bruno. (2018) *Down to earth*. politics in the new climate regime. Cambridge: Polity Press.

Latour, Bruno; Weibel, Peter (eds.). (2020) *Critical zones*: the science and politics of landing on Earth. Cambridge and Karlsruhe: MIT/ZKM.

Law, John. (1994) *Organizing modernity*. Oxford-UK and Cambridge-USA: Blackwell.

Lenton, Timothy. (2016) *Earth system science*: a very short introduction. Oxford: Oxford University Press.

Lenton, Timothy; Dutreuil, Sébastien. (2020) What exactly is the role of Gaia?. In: Latour, Bruno; Weibel, Peter (eds.). *Critical zones*: the science and politics of landing on Earth. Cambridge and Karlsruhe: MIT/ZKM.

Lohr, Virginia I. (2010) "Plants and the Individual: A Recent History." *Journal of Home & Consumer Horticulture* 1 (2–3): 225–230.

Long, Joshua and Rice, Jennifer L. (2018) From sustainable urbanism to climate urbanism. *Urban Studies* 00 (0), pp. 1-17

Luke, Timothy. (1995) On environmentality: geo-power and eco-knowledge in the discourses of contemporary environmentalism. *Cultural Critique*, n.31, p.57–81.

Luque-Ayala, Andrés. (2018) Post-development carbon. In: Luque-Ayala, Andrés; Marvin, Simon and Bulkeley, Harriet eds. *Rethinking urban transitions*: politics in the low carbon city. New York: Routledge.

Luque-Ayala, Andrés; Marvin, Simon and Bulkeley, Harriet. (2018a) Introduction. In: Luque-Ayala, Andrés; Marvin, Simon and Bulkeley, Harriet eds. *Rethinking urban transitions*: politics in the low carbon city. New York: Routledge

Luque-Ayala, Andrés; Marvin, Simon and Bulkeley, Harriet. (2018b) Rethinking urban transitions: an analytical framework. In: Luque-Ayala, Andrés; Marvin, Simon and Bulkeley, Harriet eds. *Rethinking urban transitions*: politics in the low carbon city. New York: Routledge.

Madsen, S. H. J., & Hansen, T. (2019). Cities and climate change—Examining advantages and challenges of urban climate change experiments. *European Planning Studies*, 27(2), 282–299.

Malm, Andreas. (2016) *Fossil capital*: the rise of steam power and the roots of global warming. London: Verso.

Malm, Andreas. (2021) How to blow a pipeline: learning to fight in a world on fire. London: Verso.

Maniglier, Patrice. (2020) How many Earths? The geological turn in anthropology. *The Otherwise*, n.1, 61–75.

Mann, Geoff and Wainwright, Koel. (2018) *Climate Leviathan*: a political theory of our planet future. London: Verso.

Marchart, Oliver. (2011) Democracy and minimal politics: The political difference and its consequences. *South Atlantic Quarterly* 110(4): 965–973.

Marres, Noortje. (2015) *Material participation*: technology, the environment and everyday publics. London: Palgrave Macmillan.

Marvin, S., & Rutherford, J. (2018). Controlled environments: An urban research agenda on microclimatic enclosure. *Urban Studies*, 55(6), 1143–1162. https://doi.org/10.1177/0042098018758465

McFarlane, Colin. (2011a) The city as assemblage: dwelling and urban space. *Environment and Planning D: Society and Space*, vol. 29, pages 649-671

McFarlane, Colin. (2011b). "Assemblage and Critical Urbanism." *City* 15 (2): 204–224.

Mitchell, Timothy. (2011) *Carbon democracy:* Political power in the age of oil. London: Verso.

Meerow, Sara; Newell, Joshua P.; Stults, Melissa. (2016) Defining urban resilience: A review. *Landscape and Urban Planning* 147, 38–49.

Mostafavi, Mohsen; Doherty, Gareth (eds.). (2016). *Ecological urbanism.* Cambridge: Harvard University Graduate School of Design.

Myers, Natasha. (2019) From Edenic apocalypse to gardens against Eden: plants and people in and after the Anthropocene. Hetherington, Kregg (ed.). *Infraestructures, environment, and life in the anthropocene.* Durham: Duke Press University.

Nixon, Rob. (2011) Slow Violence and the Environmentalism of the Poor. Cambridge, MA: Harvard University Press.

North, Peter and Longhurst, Noel. (2012) Grassroots localisation? The scalar potential of and limits of the 'transition' approach to climate change and resource constraint. *Urban Studies* 50 (7), pp. 1423-1438.

Nold, Christian (2018) Turning controversies into questions of design. In Marres, Noortje; Guggenheim, Michael; Wilkie, Alex. *Inventing the social.* Manchester: Mattering Press.

Oreskes, Naomi; Conway, Erik. (2010) *Merchants of doubt*: how a handful of scientists obscured the truth on issues from tobacco smoke to global warming. New York: Bloomsbury.

Pignarre, Philippe; Stengers, Isabelle. (2011) *Capitalist sorcery*: breaking the spell. New York: Palgrave and Macmillan.

Puig de la Bellacasa, Maria. (2017) *Matters of care*: speculative ethics in more than human worlds. Minneapolis: University of Minnesota Press.

Rockstrom et. al. (2009) A safe operating space for humanity, *Nature*, v. 461, set.

Rockstrom et al. (2018) Trajectories of the Earth System in the Anthropocene. *PNAS* August 14, 115 (33), pp. 8252–8259.

Sassen, Saskia. (2010) "Cities are at the Center of Our Environmental Future." *Sapiens* 2 (3): 1–8.

Sassen, Saskia. (2014) *Expulsions*: brutality and complexity in the global economy. Cambridge: Harvard University Press.

Schlosberg, David, and Romand Coles. (2016) "The New Environmentalism of Everyday Life: Sustainability, Material Flows and Movements." *Contemporary Political Theory* 15 (2): 160–181. https://doi.org/10.1057/cpt.2015.34

Schultz, Nikolaj. (2020) New climate, new class struggles. In: Latour, Bruno; Weibel, Peter (eds.). Critical zones: the science and politics of landing on Earth. Cambridge and Karlsruhe: MIT/ZKM.

Scott, Felicity (2016). *Outlaw territories*: environments of insecurity/ architectures of counterinsurgency. New York: Zone Books.

Scott, James. (1998) *Seeing like a state*: how certain schemes to improve the human condition have failed. New Haven: Yale University Press.

Simpson, Michael. (2020) Fossil urbanism: fossil fuel flows, settler colonial circulations, and the production of carbon cities, *Urban Geography*, 01–22.

Stoetzer, Bettina. (2018) "Ruderal Ecologies: Rethinking Nature, Migration, and the Urban Landscape in Berlin." *Cultural Anthropology* 33 (2): 295–323. https://doi.org/10.14506/ca33.2.09

Sloterdijk, Peter (2015). *Sfere III*: schiume. Milano: Raffaello Cortina Editore.

Star, Susan Leigh. 1999. The ethnography of infrastructure. *American Behavioral Scientist* 43: 377–391.

Star, Susan Leigh, and Karen Ruhleder. 1996. Steps toward an ecology of infrastructure: Design and access for large information spaces. *Information systems research* 7(1): 111–134.

Steffen, Will; Broadgate, Wendy; Deutsch, Lisa; Gaffney, Owen and Ludwig, Cornelia. (2015) The trajectory of the Anthropocene: the Great Acceleration. *The Anthropocene Review* Vol. 2(1) 81–98.

Steffen, Will; Katherine Richardson; Johan Rockström; Sarah E. Cornell; Ingo Fetzer; Elena M. Bennett; Reinette Biggs; Wim de Vries; et al. (2015) "Planetary Boundaries: Guiding Human Development on a Changing Planet." *Science* 347 (6223): 1259855. https://doi.org/10.1126/science.1259855

Stengers, Isabelle. (2005) Introductory notes on an ecology of practices. *Cultural Studies Review*, 11(1), 183–196.

Stengers, Isabelle. (2015a) *In catastrophic times*: resisting the coming barbarism. London: Open Humanities Press.

Stengers, Isabelle. (2015b). *Cosmopolitiche*. Roma: Luca Sossella editore.

Stengers, Isabelle. (2011) *Thinking with Whitehead*: a free and wild creation of concepts. Cambridge: Harvard University Press.

Stengers, Isabelle. (2020a) *Réactiver le sens commun*: lecture de Whitehead en temps de débâcle. Paris: Les empêcheurs de penser en rond.

Stengers, Isabelle. (2020b) We are divided. *e-flux journal*, issue 114, December, 01–06.

Stengers, Isabelle. (2021) Putting Problematization to the Test of Our Present. *Theory, Culture & Society*, Vol. 38(2) 71–92.

Stépanoff, Charles. (2024) Attachements. Enquête sur nos liens au-delà de l'humain. Paris: La Découverte.

Strathern, Marilyn. (1988) The concept of society is theoretically obsolete. In: Ingold, Tim (ed.). *Key debates in anthropology*. London and New York: Routledge.

Strathern, Marilyn. (1990). *The gender of the gift*. Problems with Women and Problems with Society in Melanesia. California: University of California Press.

Swyngedouw, Eric. (2010) Apocalypse forever? Post political populism and the spectre of climate change. *Theory, culture and Society*; vol.27(2–3): 213–232.

Szerszynski, Bronislaw and John Urry. (2010) "Changing Climates: Introduction," *Theory, Culture & Society*, 27(2-3): 1-8.

Tilly, Charles. (2008) Contentious Performances. Cambridge: Cambridge University Press.

Tsing, Anna. (2015) *The mushrooms at the end of the world*: on the possibility of life in capitalist ruins. Princeton: Princeton University Press.

Tsing, Anna Lowenhaupt; Jennifer Deger; Alder Keleman Saxena and Feifei Zhou. (2020) Feral Atlas: The More-than-Human Anthropocene. Stanford: Stanford University Press.

Turpin, Etienne. (2013) "Introduction: Who Does the Earth Think It Is, Now?" In Architecture in the Anthropocene: Encounters Among Design, Deep Time, Science and Philosophy, edited by Etienne Turpin, 7–42. Ann Arbor, MI: Open Humanities Press.

United Nations Human Settlements Programme (UN-Habitat). (2011) State of the World's Cities 2010/2011: Bridging the Urban Divide. Nairobi: UN-Habitat.

Urry, John. (2011) *Climate Change and Society*. Malden: Polity Press.

Viveiros de Castro, Eduardo. (2014) *Cannibal Metaphysics*. Minneapolis: Univocal Publishing.

Viveiros de Castro, Eduardo. (2019) On Models and Examples Engineers and Bricoleurs in the Anthropocene. *Current Anthropology*; 60:S20: S296-S308

Viveiros de Castro, Eduardo; Danowski, Déborah. (2014) *Há mundo por vir?* Ensaios sobre os medos e os fins. Florianópolis: Cultura e Barbárie/ISA.

von Schnitzler, Antina. (2016) *Democracy's Infrastructure: Techno-Politics and Protest after Apartheid*. Princeton, NJ: Princeton University Press.

Wachsmuth, David and Angelo, Hillary. (2018) Green and gray: new ideologies of nature in urban sustainability policy, *Annals of the American Association of Geographers*, 108:4, pp. 1038-1056.

Wark, McKenzie. (2017) "From Architecture to Kainotecture." e-flux Architecture, December 2017. https://www.e-flux.com/architecture/accumulation/122201/from-architecture-to-kainotecture

While, Aidan; Jonas, Andrew and Gibbs, David. (2010) From sustainable development to carbon control: eco-state restructuring and the politics of urban and regional development. *Transactions of the Institute of British Geographers*, n.35, pp. 76–93.

While, Aidan; Whitehead, Mark. (2013) Cities, Urbanisation and Climate Change. *Urban studies* 50(7) pp.1325–1331, may.

Whitehead, Mark. (2013) Neoliberal urban environmentalism and the adaptive city: towards a critical urban theory and climate change. *Urban Studies* 50(7) 1348–1367.

Yates, Luke. (2014) Rethinking prefiguration: alternatives, micropolitics and goals in social movements, *Social Movement Studies*: Journal of Social, Cultural and Political Protest: 1–20.

Zeiderman, Austin. (2016) "Adaptive Publics: Building Climate Constituencies in Bogotá." *Public Culture* 28(279): 389–413. https://doi.org/10.1215/08992363-3427499

Zetter, Roger and Morrissey, James. (2014) The environment-mobility nexus: reconceptualizing the links between environmental stress, (im)mobility and power. In: Fiddian-Qasmiyeh, Elena; Loescher, Gil; Long, Katy; Sigona, Nando. *The Oxford Handbook of Refugees and Forced Migration Studies*. Oxford University Press, New York.

Index

D. Delatin Rodrigues, *Climate Change and Political Action in Italy*, https://doi.org/10.1007/978-3-031-98771-7

FSC
www.fsc.org
MIX
Papier | Fördert gute Waldnutzung
FSC® C083411

Kolibri 360 GmbH
Ferdinand-Jühlke-Straße 7
99095 Erfurt, Deutschland
produktsicherheit@kolibri360.de